Asteroid Collision

FIFTH EDITION

ASTEROID COLLISION

Threats and Solutions

CAN BARTU H.

2024

Asteroid Collision

Can Bartu H.

All publishing rights of this book belong to CBF Publishing. Except for excerpts not exceeding one page in total for promotional purposes without the written permission of the publisher, the whole or part of the book may not be published or reproduced in any medium

CONTENTS

CONTENTS 5

CHAPTER 1 9

Introduction 9

   1.1. Solar System and Asteroids..11

   1.2. Asteroid Threat: Why It Matters? 14

   1.3 The History of Asteroid Collisions and Their Impact 19

   1.4. Asteroids in Mythology and Human Imagination 24

CHAPTER 2 33

Origin and Classification of Asteroids..33

   2.1. Formation of the Solar System and Asteroids 35

   2.2. Classification and Characteristics of Asteroids 38

   2.3. Potentially Hazardous Asteroids (PHAs) 43

   2.4 Identification and Tracking of Near-Earth Objects (NEOs)...............48

   2.5. The Yarkovsky Effect and Orbital Drift 54

   2.6. The Role of Binary Asteroids in Collision Risk 62

CHAPTER 3 69

Effects and Traces of Asteroid Collisions...69

   3.1. Gravitational Effects and Mass Extinctions.....................................71

   3.2. Impact Craters and Traces 75

   3.3 Long-Term Environmental Changes and Climate Impact................79

   3.4. Psychological Impact of Global Threats 85

   3.5. Cultural Reflections of Historical Impact Events 90

   3.6. Oceanic Impacts and Tsunami Models 95

CHAPTER 4 101

Potential Collision Scenarios 101

   4.1. Study of HighRisk Asteroids 103

   4.2. Possible Collision Points and Potential Effects.............................106

   4.3 Modeling Impact Scenarios and Probabilities...............................110

4.4. Near-Miss Events: What We've Learned...117

CHAPTER 5..123

Asteroid Threat Mitigation Methods..123

5.1. Impulse Techniques and Deflection Solutions 125

5.2. Use of Nuclear Explosions and Risks 129

5.3 Mass Evacuation Strategies and Earth Protection Plans.............................135

5.4. Role of Private Aerospace Companies in Defense 141

CHAPTER 6..147

Advanced Technologies and Space Missions 147

6.1. Space Observation and Exploration Vehicles..149

6.2. CuttingEdge Technologies and Research on Asteroids 154

6.3 Upcoming Asteroid Missions and Their Potential..159

CHAPTER 7..167

Crisis Management and Public Awareness..167

7.1. Crisis Management all through Potential Threats 169

7.2. Public Awareness and Education 174

7.3 International Cooperation and Policy Framework......................................180

7.4. The Ethics of Withholding Threat Information from the Public.................187

CHAPTER 8..193

Future Threats and Precautions 193

8.1. Impact of Advances in Science and Technology 195

8.2. ForwardLooking Predictions and Scenarios 200

8.3 The Role of AI and Machine Learning in Asteroid Detection 205

8.4. Planetary Defense within the Context of Global Security..........................212

CHAPTER 9..219

Asteroid Mining and Ethical Questions 219

9.1. Commercial Interests vs Planetary Protection...221

9.2. Legal and Political Ramifications of Resource Claims 225

CHAPTER 10..233

Conclusion 233

10.1. Summary and Evaluation 235

10.2. Recommendations for the Future...239

10.3. Promising Developments and Resources..245

CHAPTER 1

Introduction

1.1. Solar System and Asteroids

The formation of the solar device is a amazing astronomical event that took place around 4.6 billion years in the past. It all started out with a widespread, diffuse cloud of gasoline and dirt known as the solar nebula. The solar nebula consisted by and large of hydrogen and helium, along side strains of heavier factors.

Under the influence of its personal gravity and outside elements, inclusive of shockwaves from a close-by supernova or the passage of every other star, the solar nebula began to contract and crumble. As the contraction persisted, the nebula commenced to spin faster, sooner or later pulling down into a rotating disklike shape known as the protoplanetary disk.

The protoplanetary disk had a principal bulge wherein the bulk of the fabric accumulated, in the end forming the protostar that would end up the Sun. The closing fabric within the disk began to clump collectively because of collisions and gravitational attraction, forming small planetesimals. These planetesimals have been the building blocks of planets.

The manner of accretion continued because the planetesimals collided and merged to form large bodies known as protoplanets. As protoplanets grew in length, their gravity extended, attracting even extra material from the encompassing disk. The largest protoplanets sooner or later became the core of the presentday planets in our solar machine.

The early sun gadget turned into a chaotic region, with severa collisions and interactions going on among the growing planets. As a end result of these collisions, a few protoplanets grew swiftly, at the same time as others had been shattered or ejected from the solar machine altogether.

After thousands and thousands of years of this planetbuilding method, the solar device settled into its current configuration. The terrestrial planets, Mercury, Venus, Earth, and Mars, shaped in the direction of the Sun, where it turned into too warm for volatile gases like hydrogen and helium to condense. As a end result, these planets are particularly composed of rocky materials.

On the other hand, the gas giants, Jupiter, Saturn, Uranus, and Neptune, formed farther out inside the less warm areas of the protoplanetary disk. They have been in a position to accumulate significant amounts of hydrogen and helium, making them frequently composed of gasoline.

The formation and shape of the solar system stay subjects of ongoing research and exploration. By reading our sun machine's records, scientists advantage insights into planetary formation procedures and the conditions vital for life to develop. Moreover, information the solar system's shape presents a basis for evaluating and decoding other planetary systems discovered at some stage in the universe.

Asteroids are small rocky our bodies that exist at some stage in our solar gadget. They are remnants from the early degrees of its formation, regularly called the "constructing blocks" that by no means coalesced into fullfledged planets. These gadgets can range extensively in size, ranging from some meters to hundreds of kilometers in diameter.

The majority of asteroids are located within the asteroid belt, a area located among the orbits of Mars and Jupiter. Despite the famous perception that the asteroid belt is densely full of asteroids, the space between those items is giant, with a especially low density of asteroids.

Asteroids are composed of diverse materials, including rock, steel, and once in a while organic compounds. Their composition can range depending on their location inside the sun device and their unique formation history. Scientists have classified asteroids into numerous types based on their spectral characteristics, which give insights into their surface composition and beginning.

The Ctype asteroids, also referred to as carbonaceous asteroids, are the maximum commonplace kind and are rich in carbon and organic compounds. They are believed to be many of the oldest objects within the sun machine. The Stype asteroids, or silicaceous asteroids, are composed particularly of silicate minerals and are especially brighter than the Ctype asteroids. Lastly, the Mtype asteroids, or metal asteroids, are composed ordinarily of metallic and can be located both in the asteroid belt and nearEarth area.

Asteroids come in diverse shapes and sizes, starting from irregularly formed items to people with almost spherical shapes. Some asteroids have even been observed to have their very own moons, or natural satellites, that orbit round them.

One of the most substantial traits of asteroids is their capability effect risk. While maximum asteroids within the asteroid belt pose no risk to Earth, some orbits intersect with our planet's direction around the Sun. These are referred to as NearEarth Asteroids (NEAs), and they're closely monitored by using area companies and observatories to perceive any ability effect risks.

Studying asteroids offers valuable facts approximately the early solar gadget's conditions and the processes that formed its formation. It also enhances our know-how of planetary dynamics and the distribution of materials within the universe. As we keep to explore asteroids through area missions and observations, we advantage insights into the origins of our sun machine and the wider cosmos.

1.2. Asteroid Threat: Why It Matters?

Throughout the records of Earth, past asteroid collisions have had big and farreaching effects on our planet and its inhabitants. Some of the maximum fantastic affects have left a lasting impact on geological and biological evolution. Here are some key consequences of past asteroid collisions:

Mass Extinctions: Perhaps the most infamous effect of an asteroid collision is the CretaceousPaleogene (KPg) extinction occasion, which befell round sixty six million years ago. An asteroid, about 10 kilometers in diameter, struck the Yucatan Peninsula in presentday Mexico. The effect launched an good sized amount of power, leading to wildfires, tsunamis, and a worldwide "impact winter." The subsequent climate alternate and darkness resulting from dust and debris within the surroundings caused the extinction of round 75% of all species, along with the dinosaurs.

Crater Formation: Large asteroid affects depart behind impact craters on Earth's surface. These craters vary in length and may be numerous kilometers in diameter. Notable examples consist of the Chicxulub crater, created with the aid of the asteroid that triggered the KPg extinction, and the Vredefort crater in South Africa, one of the most important recognized effect systems on Earth. Impact craters offer essential geological markers and insights into Earth's history.

Climate Change: Asteroid affects can appreciably alter the Earth's weather. Upon impact, large amounts of dirt, gases, and vaporized material are ejected into the atmosphere. These particles can block sunlight and decrease the Earth's average temperature, leading to a cooling impact known as "effect wintry weather." The disruption of the worldwide weather may have giant effects on ecosystems, plant life, and climate patterns.

Habitat Destruction: Depending on the scale and area of the effect, the immediately surroundings of an asteroid strike can enjoy excessive devastation. The effect's shockwave can stage forests and structures, and the acute warmness generated can purpose wildfires. Additionally, the highvelocity effect of asteroids can create secondary results like earthquakes and landslides.

Geological Changes: Large asteroid influences can purpose great geological adjustments within the impacted vicinity. The force of the impact can cause the uplift of mountains, the formation of fault lines, and the displacement of rock layers. These geological changes can be preserved within the rock file, supplying treasured statistics for geologists analyzing Earth's records.

Mineral Deposits: Some asteroid influences can create favorable conditions for the formation of mineral deposits. The power and warmth generated in the course of the effect can mobilize metals and minerals, leading to their attention and eventual deposition. Impactrelated mineral deposits have financial importance and are vital for numerous industries.

Studying the results of past asteroid collisions on Earth is vital for know-how our planet's records and evolution. It additionally highlights the potential dangers posed through destiny asteroid impacts and emphasizes the importance of developing techniques for asteroid effect mitigation and early detection. As we keep to discover the solar machine, know-how of past asteroid collisions provides treasured insights into planetary dynamics and the potential impact hazards faced through all celestial bodies.

Potential destiny threats from asteroids continue to be a subject of situation and energetic research. The effect danger is a real and steady presence, and scientists and area corporations global are actively engaged in efforts to discover and check capacity dangers. Here are a few key factors of potential destiny asteroid threats:

Detection and Tracking: The first line of protection against ability asteroid threats is the detection and monitoring of NearEarth Asteroids (NEAs). Thanks to advances in era and committed space telescopes, astronomers continuously reveal the skies to become aware of any asteroids whose orbits come near Earth. This early detection is essential for assessing capability risks and developing suitable mitigation techniques.

Orbital Analysis: Once an asteroid is detected, scientists conduct unique orbital analysis to determine its future trajectory. Understanding an asteroid's orbit enables predict whether it poses a chance of colliding with Earth in the coming years, a long time, or maybe centuries. This records allows for well timed warnings and planning.

Risk Assessment: Determining the capacity impact chance of an asteroid entails considering various factors, together with its size, speed, composition, and effect attitude. Large asteroids with excessive velocities and dense compositions can purpose greater big damage upon impact. Scientists use state-of-the-art models and simulations to assess the ability results of an impact.

Mitigation Strategies: If an asteroid is identified as posing a ability threat, diverse mitigation strategies can be taken into consideration. These consist of techniques to regulate the asteroid's trajectory or to disrupt it to prevent a collision with Earth. Techniques along with gravity tractors, kinetic impactors, and sun sails were proposed and studied as possible way of deflecting an asteroid faraway from its collision route.

International Collaboration: Asteroid effect threats are worldwide issues, and addressing them calls for international collaboration and cooperation. Space organizations, studies institutions, and governments from round the sector proportion data and coordinate efforts to track and have a look at NearEarth Asteroids. Collaborative projects enhance our collective ability to reply efficaciously to ability threats.

Public Awareness: Raising public focus about asteroid effect threats is important. By instructing the general public approximately the technological know-how behind asteroid detection, impact modeling, and capacity consequences, individuals can better recognize the importance of this difficulty. Public consciousness additionally helps foster aid for area missions aimed toward asteroid research and planetary protection.

While the chance of a catastrophic asteroid impact is rather low, the ability outcomes are immense. The ongoing efforts to monitor, detect, and assess capacity destiny threats are important for our planet's protection and the preservation of lifestyles on Earth. By staying vigilant and proactive in our approach to asteroid impact threats, we can paintings collectively to protect our planet from this cosmic chance.

1.3 The History of Asteroid Collisions and Their Impact

The history of asteroid collisions is not just a story of celestial bodies interacting within the vastness of space; it is a narrative that stretches back billions of years, shaping the Earth and, in many cases, the very course of life itself. These impacts have had profound effects on the Earth's geology, climate, and biosphere, and continue to be a key focus of scientific research. Understanding the history of asteroid collisions provides crucial insights into the potential future threats that Earth faces from these space objects.

The first recorded asteroid collisions occurred in the early history of the solar system, around 4.5 billion years ago, when the planets and smaller bodies were in the process of formation. This period, often referred to as the Late Heavy Bombardment, was characterized by intense meteorite and asteroid activity, which significantly affected the young Earth. During this time, the Earth's surface was bombarded by a vast number of space objects, from small meteoroids to large asteroids.

These collisions were instrumental in shaping the Earth's surface, with the creation of large impact basins and craters. The scars left on the Earth's surface during this period are still visible today, though many have been eroded or covered by subsequent geological activity. It is believed that these early impacts played a significant role in the formation of the Earth's atmosphere and oceans, possibly even contributing to the chemical ingredients necessary for the development of life.

One of the most famous asteroid collisions in Earth's history occurred approximately 66 million years ago, during the Cretaceous-Paleogene (K-T) boundary event. This event marked the end of the Cretaceous period and the extinction of about 75% of Earth's species, including the dinosaurs. The cause of this mass extinction has been attributed to the impact of a large asteroid, about 10 kilometers in diameter, which struck what is now the Yucatán Peninsula in Mexico. The impact created the Chicxulub crater, which is over 150 kilometers wide and more than 20 kilometers deep.

The immediate effects of the asteroid collision were catastrophic. The explosion released an enormous amount of energy, equivalent to billions of atomic bombs, causing massive fires, tsunamis, and a global dust cloud that blocked sunlight for months. This led to a dramatic drop in temperatures and disrupted photosynthesis, precipitating a "nuclear winter" effect. The long-term environmental changes, including the cooling of the planet and the disruption of ecosystems, were largely responsible for the extinction event that followed.

The K-T extinction event serves as a stark reminder of the potential dangers that asteroid collisions pose to life on Earth. While it is widely regarded as the most significant impact event in Earth's history, it is not an isolated incident. Over the past few billion years, numerous asteroid collisions have had lasting effects on the planet, often triggering dramatic climate changes, mass extinctions, and major geological events.

In addition to the K-T event, there have been many other asteroid collisions that have had significant impacts on Earth's history. While these events may not have been as large or as globally catastrophic, they still played a critical role in shaping the planet's geological and biological history.

The Tunguska Event (1908): One of the most recent asteroid impacts occurred in Siberia, Russia, when an asteroid or comet exploded in the atmosphere, releasing energy equivalent to a 15-megaton bomb. The explosion flattened over 2,000 square kilometers of forest, but because the asteroid disintegrated before hitting the ground, there were no direct casualties. The Tunguska event remains a key example of how asteroid impacts can cause massive destruction even without hitting the Earth's surface.

The Barringer Crater (Meteor Crater) in Arizona: This well-preserved impact site, created by a nickel-iron meteorite approximately 50,000 years ago, offers scientists valuable insights into the forces at play during an asteroid collision. The crater is nearly 1.5 kilometers in diameter and about 170 meters deep, providing a clear record of the explosive forces unleashed during the impact.

The Vredefort Crater (South Africa): The Vredefort crater, located in South Africa, is one of the largest and oldest known impact craters on Earth, with a diameter of around 300 kilometers. It was formed by the impact of an asteroid around 2 billion years ago. The Vredefort impact is believed to have had significant geological and climatic consequences at the time, including potentially influencing the evolution of life on Earth.

Asteroid impacts have not only altered Earth's surface but have also had a profound effect on the planet's climate and the course of biological evolution. Large impacts can trigger dramatic climate changes, ranging from global cooling due to dust clouds blocking sunlight to rapid warming due to the greenhouse gases released by fires and other processes. These climate shifts, in turn, can drive evolutionary changes, either creating new ecological niches or wiping out entire species.

The K-T extinction event, for instance, is believed to have caused a massive disruption in the Earth's ecosystems, paving the way for the rise of mammals and, eventually, the dominance of human beings. Had the asteroid impact not occurred, the dinosaurs may have continued to reign as the dominant species on Earth, possibly altering the evolutionary path of life altogether.

Other, smaller impacts have contributed to the shaping of the planet's climate, driving shifts between ice ages and warmer periods. These events have played a role in the migration of species, the rise of new biological forms, and the extinction of others. As such, asteroid impacts have been a key factor in the Earth's ongoing geological and biological evolution.

Today, the threat of asteroid impacts is no longer a matter of historical interest; it is a modern-day concern. While large asteroid collisions are rare, the potential consequences of such an event are so catastrophic that they warrant serious attention. Advances in space observation and technology have enabled scientists to better track near-Earth objects (NEOs) and assess the risks they pose. The identification of potentially hazardous asteroids (PHAs) and their tracking is now a priority for space agencies around the world.

Several global initiatives are underway to develop asteroid deflection and impact mitigation strategies. These efforts are designed to protect Earth from the catastrophic effects of asteroid collisions, should one be detected as a potential threat. The history of asteroid collisions serves as a powerful reminder of the importance of these efforts and the potential dangers that lie ahead.

The history of asteroid collisions and their impact on Earth is a testament to the immense power of these celestial objects. From the early bombardment of the solar system to the mass extinctions triggered by major impact events, asteroid collisions have played a significant role in shaping the history of our planet. Understanding these events is crucial not only for appreciating the Earth's past but also for preparing for the potential threats that may arise in the future.

As technology advances and our understanding of asteroid dynamics improves, the ability to prevent or mitigate the effects of future asteroid impacts becomes increasingly possible. The study of past asteroid collisions, combined with modern scientific research, provides valuable lessons in how to protect the Earth and ensure the survival of life in the face of these cosmic threats. The history of asteroid impacts is not just a reminder of past catastrophes but also a guide to safeguarding our future.

1.4. Asteroids in Mythology and Human Imagination

From the earliest days of human awareness, celestial items have served as mirrors of our fears, aspirations, and mysteries. Among these, asteroids—even though now not seen to the naked eye until the arrival of telescopes—have found symbolic illustration in historical myths, legends, and the evolving tapestry of human creativeness. Though historical cultures couldn't distinguish asteroids from stars or meteors, their conceptual shadows loomed large in metaphors of destruction, divine wrath, rebirth, and cosmic order. In modern-day instances, with the rise of medical observation, asteroids have continued to ignite collective creativeness, serving as amazing symbols in literature, film, and philosophical inquiry.

In the mythologies of many cultures, the sky become now not a passive dome however an active, often sentient space. Celestial phenomena have been imbued with that means, and any disruption—such as a fireball streaking through the heavens—became interpreted as a message or omen. While these had been probably meteorites or comets, the archetypal role they played mirrors the issues we now partner with asteroids: unpredictable traffic from the heavens with the energy to create or break.

In Mesopotamian cosmology, the heavens have been dominated by gods whose tempers formed the earth. Enlil, the storm god, turned into acknowledged to cast down destruction whilst humanity displeased the divine order. While the texts in no way mention asteroids by call, the belief of fiery items solid from the heavens unearths near parallel in later Judeo-Christian and Islamic traditions. The "burning stones" of divine judgment, as defined in the Book of Genesis in the course of the destruction of Sodom and Gomorrah, or within the Qur'an at some stage in the stories of Aad and Thamud, fit the archetype of celestial punishment—a role asteroids have come to embody in cutting-edge catastrophe psychology.

Greek mythology provided a greater based pantheon, but it too featured narratives that echo the suddenness and strength of asteroid-like activities. Phaethon, the son of Helios, who tried to pressure the chariot of the solar and misplaced manipulate, scorched the earth before being struck down by way of Zeus. Though fantasy, this tale evokes the photo of a celestial body veering off its direction, becoming a danger to Earth, and requiring divine intervention. In another thread, the Titans—primordial forces defeated by means of the Olympians—had been regularly buried underneath mountains, their struggles ensuing in earthquakes and volcanic eruptions. The topic of ancient, huge forces hidden however not long gone speaks subtly to how we apprehend the asteroid belt: remnants of a damaged global, orbiting nevertheless, threatening silently.

The Norse sagas are replete with images of fireplace and ice falling from the skies all through Ragnarök, the give up-time apocalypse. The celestial wolves Sköll and Hati are fated to consume the solar and moon, plunging the cosmos into chaos. In the ensuing darkness, stars fall and the sky itself splits open. These catastrophic snap shots align uncannily with modern-day conceptions of asteroid effect scenarios—the blotting of the sun with the aid of ejecta, the impact winter, the fall apart of order. Though the ancients did not have our astronomical knowledge, their myths captured the deep, mental resonance of cosmic events.

In Mesoamerican mythology, especially the various Maya and Aztecs, the sky was each sacred and perilous. The god Tezcatlipoca, frequently depicted as a jaguar or a smoking mirror, was associated with chaos, night time, and destruction. The Aztecs believed in cycles of solar a while, every finishing in cataclysm—wind, fireplace, flood, or jaguars. Some researchers have drawn symbolic hyperlinks among those catastrophic endings and the sort of global devastation that would be due to huge asteroid affects. The notion of cyclical destruction and rebirth, written into their cosmology, mirrors the geological cycles of mass extinction and evolutionary renewal prompted through affects.

In Hindu mythology, the tale of Shiva—the destroyer and regenerator—can be visible as symbolic of cosmic methods akin to asteroid influences. Shiva's dance, the Tandava, is both the destruction of the universe and the prelude to its rebirth. While no unique celestial item is called, the themes of sudden, violent transformation leading to renewal resonate with how cutting-edge planetary scientists understand the dual nature of asteroid impacts: each a purpose of mass extinction and a catalyst for evolutionary leaps.

The Abrahamic traditions further describe celestial violence as divine intervention. The Book of Revelation speaks of "a brilliant star, blazing like a torch, falling from the sky…

Asteroid Collision | **27**

poisoning a third of the waters." Named Wormwood, this prophecy has stimulated limitless modern depictions of asteroid-precipitated Armageddon. In Islamic way of life, the trumpet of Israfil is said to signal the end of days, followed by using the sky splitting aside and stars scattering. These metaphors, at the same time as theological, evoke the physical approaches we now recognize to be brought about through massive influences: fragmentation, shockwaves, chemical modifications, and darkness.

As human know-how of the cosmos evolved, especially via the Enlightenment and the clinical revolution, mythology began to provide way to imagination fashioned by reason. Yet the romantic and symbolic electricity of celestial threat persisted. In 1801, the discovery of Ceres, the first asteroid, rekindled interest in the concept of a broken planet—probably the remains of a 5th terrestrial world once orbiting the Sun among Mars and Jupiter. This notion gave asteroids an added layer of thriller: no longer simply celestial wanderers, but the shattered bones of a cosmic tragedy.

In the nineteenth and twentieth centuries, literature embraced asteroids as metaphors for fate, insanity, and the sublime. Jules Verne and H.G. Wells speculated approximately impacts and the existential danger posed via area. Wells' The Star, published in 1897, describes a celestial item on a collision path with Earth, causing floods, volcanoes, and panic. The tale faucets into the deep mental worry of a hazard too considerable to govern, too remote to forestall—an concept that modern-day impact technological know-how has only reinforced.

In cinema, the asteroid took on new symbolic dimensions. The Cold War era converted cosmic threats into political allegories. In the 1998 movies Armageddon and Deep Impact, asteroids have become stand-ins for uncontrollable international crises, whether or not nuclear warfare, climate catastrophe, or pandemic. They supplied a degree on which humanity should confront its mortality, technological limits, and potential for collective motion. Even extra abstractly, they embodied randomness—the concept that civilization could stop now not thru hubris or sin, however definitely due to the fact a rock hit the wrong place at the wrong time.

The human creativeness has extensively utilized asteroids as a canvas for wish and growth. In technological know-how fiction narratives, asteroids were reimagined as houses, fortresses, and vessels. Arthur C. Clarke's Rendezvous with Rama recommendations at a giant asteroid-fashioned ufo, at the same time as Kim Stanley Robinson's 2312 envisions a destiny where hollowed-out asteroids serve as interplanetary arks. These visions invert the myth of destruction and reframe asteroids as lifeboats, symbols of variation and survival.

The symbolic evolution of asteroids from mythic punishment to modern salvation reflects broader shifts in human cosmology. In in advance eras, the sky became a realm of gods, and any anomaly turned into a signal of divine desire or wrath. In modernity, area is both a chance and a frontier. Asteroids, once sellers of divine destruction, are actually objectives for mining, colonization, and planetary protection. This transformation marks a profound shift in how humanity pertains to the unknown.

Yet in spite of the scientific demystification, asteroids maintain their mythic fee. The 2013 Chelyabinsk meteor reminded the world that the sky is not empty. Its explosion over Russia injured hundreds and illuminated the fragility of our global infrastructure. The psychological impact of such activities—the surge in tension, the deluge of conspiracy theories, the renewed interest in Near-Earth Objects—echoes historical responses. The sky stays a supply of awe and worry.

Asteroids additionally feature prominently in present day artwork and philosophy. Visual artists use effect imagery to explore themes of entropy, chaos, and renewal. Philosophers draw on the idea of asteroid influences to speak about contingency in history—the idea test of what if the Chicxulub impact had overlooked, and the dinosaurs had by no means gone extinct. Would mammals have risen? Would people exist? In this context, the asteroid is not just an item but a fulcrum of trade realities.

The asteroid additionally symbolizes scale—each temporal and spatial. The rocks orbiting among Mars and Jupiter are relics of a time billions of years beyond. They carry with them the memory of the sun system's violent formation, the scars of ancient collisions, and the minerals of planetary cores. Holding an asteroid sample in a laboratory is corresponding to touching the bones of the cosmos. In this manner, the asteroid becomes a metaphor for time itself—deep, silent, and sizeable.

In mental phrases, asteroids frequently characterize the unpredictable. In Jungian evaluation, they are able to constitute archetypes of surprising trade or disagreement with the unconscious. Their orbits, regularly elliptical and difficult to song, replicate the chaotic forces in human existence that appear to emerge with out warning. The perception of a "hidden chance" aligns with fears buried within the collective unconscious—an tension that some thing historical and effective nevertheless watches from the void.

Cultural traditions preserve to combine asteroids into their symbolic vocabulary. In astrology, newer interpretations have began to encompass recognized asteroids like Chiron and Vesta, assigning them roles in private and collective identification. While not clinical, those practices display the long-lasting human impulse to discover that means inside the heavens. Even nowadays, when we call asteroids—after gods, lovers, cities, and poets—we're persevering with a mythic lifestyle.

Asteroids occupy a unique space in human imagination. They are actual, measurable, and dangerous. But they may be also symbols—of fate, trade, memory, and potential. Whether feared as harbingers of doom or envisioned as stepping stones to the stars, they reflect the duality on the coronary heart of human life: fragility and resilience, destruction and renewal, fable and motive. The tales we tell about them—ancient or modern-day—display as lots about ourselves as they do approximately the cosmos.

CHAPTER 2

Origin and Classification of Asteroids

2.1. Formation of the Solar System and Asteroids

The Nebular Theory is a broadly well-known cause of the formation of the sun machine and different planetary systems within the universe. Proposed by means of Immanuel Kant and similarly advanced through PierreSimon Laplace in the 18th century, the concept suggests that our solar gadget originated from a huge cloud of gasoline and dirt referred to as the "solar nebula."

According to the Nebular Theory, the sun nebula become a rotating, flattened disk with the majority of its mass focused at the middle. This important area, known as the "protoplanetary disk" or "protoplanetary nebula," contained most of the material that could later shape the Sun.

As the solar nebula gotten smaller under its own gravity, it started to spin faster because of the conservation of angular momentum. This spinning movement brought about the nebula to flatten right into a disk form, with a bulging middle wherein the destiny Sun fashioned.

Over time, the protoplanetary disk persevered to chill and condense, leading to the accumulation of strong particles and dirt grains referred to as "planetesimals." Planetesimals are small bodies that function the constructing blocks for planets.

The protoplanetary disk also skilled a few diploma of turbulence and turbulence in its early tiers, leading to the migration of cloth inside the disk and the redistribution of mass. These processes contributed to the growth of planetesimals thru collisions and accretion.

As planetesimals endured to develop, they merged to form large bodies called "protoplanets." The protoplanets continued to accrete surrounding material till they reached sufficient mass and gravity to clear their orbits of particles, turning into the planets we recognise nowadays.

The Nebular Theory not handiest explains the formation of planets but additionally bills for the presence of other celestial bodies in the solar system. It suggests that asteroids, comets, and other minor gadgets also originated from the same protoplanetary disk however did no longer accrete into fullfledged planets.

While the Nebular Theory presents a comprehensive cause of the solar machine's formation, ongoing research and space missions, which includes NASA's James Webb Space Telescope, hold to deepen our expertise of this charming and dynamic process. By reading the formation of our sun gadget, scientists gain valuable insights into the delivery of planetary structures during the universe.

Protoplanets and Asteroid Belts

During the early levels of the sun machine's formation, because the protoplanetary disk step by step cooled and solidified, planetesimals persevered to collide and accrete, main to the formation of protoplanets. Protoplanets are enormously massive and planetlike bodies that have now not yet reached the dimensions and traits of fully advanced planets.

In the protoplanetary disk, some protoplanets grew to a significant length, but they did now not grow to be big enough to clean their orbits of surrounding particles and planetesimals. These items, ranging in size from some meters to numerous hundred kilometers, shaped what we now know as the asteroid belts.

The asteroid belts are areas inside the sun system where a substantial range of asteroids orbit the Sun. The maximum fashionable and full-size asteroid belt is placed between the orbits of Mars and Jupiter. It is often known as the "Main Asteroid Belt."

The Main Asteroid Belt carries a various populace of asteroids, varying in size, composition, and form. Most asteroids in the Main Belt are composed of rock and metallic, and they're categorised into specific groups based totally on their spectral characteristics. The Ctype asteroids (carbonaceous) are the most ample, observed via the Stype asteroids (silicaceous), and the Mtype asteroids (steel).

The gravitational have an effect on of Jupiter, the largest planet in our sun device, plays a important function in the shape and dynamics of the Main Asteroid Belt. Jupiter's gravity allows maintain gaps and clean regions within the belt, referred to as "Kirkwood gaps," in which asteroids with certain orbital resonances are much less possibly to be discovered.

In addition to the Main Asteroid Belt, there are different smaller asteroid belts or companies of asteroids within the solar gadget. For instance, the Trojans are two groups of asteroids that percentage the equal orbit as Jupiter, main or trailing the planet by means of about 60 stages.

Asteroid belts play an vital role in know-how the early history of the solar gadget. They incorporate valuable statistics approximately the constructing blocks of planets and the techniques of planetary formation and migration. Asteroids within the Main Belt had been studied via space missions, consisting of NASA's Dawn task, which visited the 2 largest asteroids within the Main Belt, Vesta and Ceres.

Although the asteroid belts include an abundance of asteroids, they're additionally vast areas of space, with huge distances among man or woman items. As a result, the possibility of spacecraft encountering asteroids at some point of space missions is distinctly low, and missions are cautiously planned to ensure secure passage thru these regions. Studying the asteroids in those belts keeps to offer valuable insights into the origins and evolution of our solar machine.

2.2. Classification and Characteristics of Asteroids

Asteroids in the solar machine showcase a huge range of compositions, and scientists have classified them into various lessons based on their spectral characteristics, which replicate the styles of materials present on their surfaces. The 3 number one training of asteroids are Ctype, Stype, and Mtype, each representing distinct organizations based on their compositions:

Ctype (Carbonaceous) Asteroids: Ctype asteroids are the maximum commonplace kind, constituting about 75% of all known asteroids. These asteroids are wealthy in carbon and are believed to be among the oldest items in the sun machine. They originated from the outer areas of the protoplanetary disk, where lower temperatures allowed risky compounds like water, methane, and ammonia to condense.

Ctype asteroids are darkish and have low albedo, which means they replicate best a small fraction of the daylight that falls on them. Their spectra show absorption capabilities associated with the presence of minerals including clay and carbon compounds. Ctype asteroids are usually located within the outer regions of the Main Asteroid Belt and past, as well as in different regions of the solar machine.

Stype (Silicaceous) Asteroids: Stype asteroids make up about 17% of all regarded asteroids. They are composed often of silicate minerals, consisting of pyroxenes and olivines, that are not unusual rockforming minerals determined on Earth. Stype asteroids are brighter and feature better albedo as compared to Ctype asteroids.

These asteroids probably originated from the internal regions of the protoplanetary disk, wherein better temperatures avoided risky compounds from condensing. Stype asteroids are normally determined within the internal areas of the Main Asteroid Belt, closer to Mars and Jupiter.

Mtype (Metallic) Asteroids: Mtype asteroids constitute approximately 10% of all recognised asteroids. They are composed mainly of metals, specifically nickel and iron. Mtype asteroids are distinctly vivid and feature excessive albedo because of the metallic nature in their surfaces.

These asteroids are thought to have originated from the metal center of huge, differentiated determine bodies that experienced considerable heating and melting throughout the early ranges of the sun device. Many Mtype asteroids are observed in the center regions of the Main Asteroid Belt.

Apart from the primary Ctype, Stype, and Mtype asteroids, there are different, much less common classes of asteroids with specific spectral traits. For example, Xtype asteroids showcase a completely unique absorption function related to olivinerich surfaces. Dtype (Dark) asteroids are much like Ctype asteroids but have even decrease albedo. Vtype (Vestoid) asteroids are a subgroup of Vesta, the secondlargest item inside the Main Asteroid Belt, and are believed to have originated from Vesta's huge impact craters.

Understanding the composition and distribution of asteroid training gives precious facts about the history and evolution of the sun device. By reading those special sorts of asteroids, scientists gain insights into the situations and strategies that shaped the early protoplanetary disk and the diverse materials present throughout the solar device's formation.

Size Distribution and Diameters

Asteroids within the sun device come in a huge range of sizes, from small pebblelike gadgets to massive our bodies numerous hundred kilometers in diameter. Understanding the size distribution of asteroids is important for reading their population and the general traits of the asteroid population. Here are a few key elements of the size distribution and diameters of asteroids:

Size Distribution: The size distribution of asteroids follows a powerlaw distribution, also referred to as the "sizefrequency distribution." This distribution way that the variety of asteroids will increase as their size decreases. In other words, there are many greater small asteroids than big ones.

The majority of asteroids within the solar system are especially small, with diameters ranging from a few meters to three kilometers. These small asteroids make up the full-size majority of the population but have constrained ordinary mass. As the size of asteroids increases, their numbers lower unexpectedly. Large asteroids, with diameters exceeding a hundred kilometers, are surprisingly uncommon.

Diameters of Asteroids: Asteroids range substantially in size, and their diameters may be measured in meters, kilometers, or maybe hundreds of kilometers. The diameters of asteroids are envisioned the usage of numerous strategies, along with direct measurements from spacecraft flybys or telescopic observations combined with brightness measurements and assumptions approximately the asteroids' albedo (reflectivity).

The smallest asteroids, referred to as "microasteroids" or "boulders," have diameters starting from some centimeters to numerous meters. These small gadgets are tough to detect from Earth and are frequently best located while they arrive very close to our planet.

Small asteroids, also known as "planetesimals," have diameters inside the range of a few meters to three kilometers. These are the building blocks from which larger asteroids and protoplanets fashioned during the early degrees of the solar device.

Intermediatesized asteroids have diameters within the variety of numerous kilometers to tens of kilometers. Many of the asteroids located inside the Main Asteroid Belt fall into this length category.

The largest asteroids, also referred to as "planetsized" or "dwarf planet" asteroids, will have diameters exceeding 100 kilometers. Ceres, the biggest item inside the Main Asteroid Belt, is classified as a dwarf planet and has a diameter of approximately 940 kilometers.

It's crucial to observe that the size distribution of asteroids is constantly evolving due to diverse techniques along with collisions, fragmentation, and gravitational interactions with planets. Over time, collisions can wreck larger asteroids into smaller fragments, while gravitational interactions can regulate the orbits and distributions of asteroids within the solar machine.

Studying the size distribution and diameters of asteroids facilitates scientists benefit insights into the origins and evolution of those items. By knowledge their sizes and distribution, researchers can higher determine the capacity effect risks they pose to Earth and other planets and explore possibilities for destiny asteroid missions and area exploration.

2.3. Potentially Hazardous Asteroids (PHAs)

Potentially Hazardous Asteroids (PHAs) are asteroids that have orbits that deliver them close to Earth and feature the potential to pose a threat of collision with our planet. Identifying and reading PHAs is vital for assessing and mitigating the ability risks of asteroid impacts. Here are the key components of the identification and characteristics of PHAs:

Close Approaches to Earth: PHAs are described primarily based on their Minimum Orbit Intersection Distance (MOID) with Earth. The MOID is the closest distance among the orbit of an asteroid and the orbit of Earth. Asteroids with an MOID of zero.05 astronomical units (AU) or less (1 AU is the average distance among the Earth and the Sun, approximately 93 million miles or a hundred and fifty million kilometers) are taken into consideration PHAs.

Size and Impact Energy: PHAs can range substantially in size, starting from a few meters to numerous kilometers in diameter. The ability impact power of a PHA is determined by its size, pace, and density. Larger PHAs with better velocities should reason more vast harm upon effect.

Orbital Characteristics: PHAs have orbits that convey them close to Earth's orbit. Some PHAs may have orbits that intersect or come very near our planet's orbit. These close processes growth the chance of a destiny collision with Earth, even though the actual possibility of effect depends on different factors, including the asteroid's orbit and uncertainties in its future trajectory.

Observational Data: Astronomers use telescopic observations and radar imaging to collect information approximately PHAs. By tracking their positions through the years, scientists can decide their orbits and predict their future positions. This information allows for precise calculations of the asteroid's MOID and capacity effect possibilities.

Risk Assessment: Assessing the danger posed with the aid of a PHA entails comparing the chance of effect and the ability effects if a collision have been to arise. The Torino Scale and the Palermo Technical Impact Hazard Scale are systems used to assess the impact risk related to PHAs.

Mitigation Strategies: In the event that a PHA is recognized as posing a sizable impact threat, various mitigation techniques can be considered. These strategies goal to modify the asteroid's trajectory to keep away from a collision with Earth. As mentioned earlier, techniques inclusive of gravity tractors, kinetic impactors, and sun sails were proposed as capacity method of asteroid deflection.

Monitoring and Early Warning: Continuous tracking of PHAs is important for presenting early caution of capacity impact threats. Space companies and observatories international are engaged in ongoing efforts to locate and song PHAs to improve our know-how of their orbits and to discover any capacity hazards nicely in advance.

Studying PHAs enables us higher recognize the capability effect risks confronted through our planet and aids in growing strategies to mitigate these dangers. By figuring out and characterizing PHAs, scientists contribute to our ordinary planetary defense efforts and decorate our preparedness for any potential future asteroid threats.

Observing and monitoring Potentially Hazardous Asteroids (PHAs) is crucial for accurately figuring out their orbits, predicting their future positions, and assessing their capability impact dangers. Astronomers use diverse observational and tracking methods to acquire facts approximately PHAs. Here are the key techniques used for observing and monitoring PHAs:

Optical Telescopes: Optical telescopes are the maximum common instruments used for staring at asteroids, consisting of PHAs. These telescopes stumble on visible light and accumulate pics of the night time sky. Astronomers use CCD (ChargeCoupled Device) cameras to capture pics of asteroids over time. By evaluating photos taken on distinctive nights, scientists can become aware of the shifting points of mild corresponding to asteroids and determine their positions inside the sky.

Radar Imaging: Radar imaging is a effective technique used to have a look at asteroids, particularly those which might be close to Earth. Radar observations offer highresolution pix and permit scientists to directly degree the shape, size, rotation, and floor features of PHAs. This technique is especially useful for refining the orbits of asteroids and predicting their trajectories with extra precision.

Infrared Telescopes: Infrared telescopes are designed to stumble on infrared radiation emitted by way of items in space. Asteroids, such as PHAs, emit infrared radiation due to their thermal warmness. Infrared observations assist determine the asteroids' temperatures, sizes, and surface residences, even when they are not illuminated by way of the Sun.

SpaceBased Telescopes: Spacebased telescopes, which include the Hubble Space Telescope, Spitzer Space Telescope, and the imminent James Webb Space Telescope, offer unique advantages for staring at asteroids. They operate above the

Earth's atmosphere, which reduces atmospheric distortion and lets in for clearer and extra detailed observations.

Astrometry and Orbit Determination: Astrometry is the best dimension of the positions and motions of celestial objects. Observations from multiple telescopes are mixed to determine the positions of asteroids correctly. Astrometric facts, such as the apparent brightness, function, and motion of PHAs, are used to calculate their orbits and future trajectories.

Lightcurve Analysis: The lightcurve of an asteroid refers to its brightness version over time as it rotates. Lightcurve analysis affords treasured statistics approximately an asteroid's rotation length, shape, and on occasion the presence of multiple objects (binary asteroids) or abnormal shapes. Studying lightcurves enables astronomers signify PHAs and understand their physical residences.

NearEarth Object (NEO) Surveys: Various groundbased observatories and devoted surveys cognizance on coming across and monitoring NEOs, inclusive of PHAs. These surveys systematically scan the sky to become aware of formerly unknown asteroids and display acknowledged ones to refine their orbits.

Continuous and systematic observations the use of these methods assist improve our understanding of PHAs, their orbits, and ability effect risks. By combining statistics from numerous observational techniques, astronomers can appropriately are expecting the future positions of PHAs and offer early caution if any pose a great impact danger to Earth. The observations and monitoring of PHAs are crucial components of our planetary defense efforts and make contributions to global efforts to enhance our preparedness for any ability asteroid influences.

2.4 Identification and Tracking of Near-Earth Objects (NEOs)

The identification and tracking of Near-Earth Objects (NEOs) are crucial to understanding the potential threats these objects pose to Earth and our global security. NEOs, which include asteroids and comets that come close to Earth's orbit, are considered significant due to their proximity and potential for impact. Over the past few decades, technological advancements and international cooperation have enabled significant progress in identifying and monitoring these objects, but challenges remain in fully assessing the risks they pose.

A Near-Earth Object is any asteroid or comet whose orbit brings it within 1.3 astronomical units (AU) of the Earth. An astronomical unit (AU) is the average distance between the Earth and the Sun, approximately 150 million kilometers. NEOs can be classified into two primary categories:

Near-Earth Asteroids (NEAs): These are asteroids that orbit the Sun in close proximity to Earth's orbit. They are further classified based on their orbits into three groups:

Atira (or Apohele) Asteroids: These asteroids have orbits entirely within Earth's orbit, crossing between the Earth and the Sun.

Amor Asteroids: These are asteroids whose orbits are outside Earth's orbit but still pass relatively close to the Earth.

Apollo and Aten Asteroids: These asteroids have orbits that cross Earth's path. Apollo asteroids have a semi-major axis greater than Earth's, while Aten asteroids have a semi-major axis smaller than Earth's.

Near-Earth Comets (NECs): These are comets whose orbits bring them close to the Earth. While their trajectories are typically more elongated and eccentric than those of asteroids, their potential to impact the Earth remains significant.

Tracking these objects is essential because even small changes in their orbits, caused by gravitational interactions or other factors, could increase the likelihood of a collision with Earth. For this reason, space agencies and observatories around the world have developed advanced techniques to detect and track NEOs.

Over the years, various methods have been employed to identify and track NEOs, ranging from ground-based telescopes to space missions. Some of the most effective techniques include:

The most common method of detecting NEOs is through ground-based telescopes. These observatories, often located in high-altitude areas to reduce atmospheric interference, scan the sky for objects moving relative to the background stars. NEOs are detected because they appear as moving points of light, unlike the distant stars that remain stationary.

The Pan-STARRS (Panoramic Survey Telescope and Rapid Response System), located in Hawaii, is one of the leading observatories for NEO detection. Pan-STARRS uses wide-field imaging and advanced software to detect moving objects in the night sky. The system is capable of scanning large areas of the sky and can quickly spot new NEOs as they cross Earth's orbit.

The Catalina Sky Survey in Arizona is another significant observatory that specializes in detecting and tracking NEOs. This survey has played a key role in discovering many new asteroids and monitoring their orbits over time.

The Mt. Lemmon Survey in Arizona also contributes significantly to the identification of NEOs, and its work is part of NASA's efforts to improve the detection and tracking of hazardous objects.

These telescopes use a combination of photographic surveys and automated systems that scan large portions of the sky, identify moving objects, and catalog their positions. Once an object is detected, its orbit is calculated, and it is classified as a potential NEO if its orbit brings it close to Earth.

In addition to ground-based observations, space telescopes have become an important tool for detecting NEOs.

Space-based observatories are particularly advantageous because they can observe NEOs in infrared wavelengths, which are less affected by Earth's atmosphere. Infrared observations help scientists identify asteroids that might be difficult to detect in visible light, as asteroids emit heat that can be detected in the infrared spectrum.

One key space-based observatory dedicated to NEO detection is NEOWISE, a mission that repurposed NASA's Wide-field Infrared Survey Explorer (WISE) satellite. NEOWISE scans the sky in infrared light to identify both large and small NEOs. It has been instrumental in finding objects that are not visible through optical telescopes.

Radar observations are another essential tool for tracking NEOs, providing highly accurate measurements of an object's position, velocity, and even surface characteristics. Radar systems bounce radio waves off NEOs, and by analyzing the reflected signals, astronomers can determine precise details about their size, shape, and motion.

The Arecibo Observatory in Puerto Rico was one of the most important radar facilities for NEO tracking, but after its collapse in late 2020, efforts have shifted to other facilities, such as the Goldstone Solar System Radar in California. Radar observations allow scientists to refine the orbits of NEOs and predict their future trajectories with great precision, which is critical for assessing potential collision risks.

Once a NEO is detected, its orbit must be tracked over time to determine whether it poses a threat to Earth. Small asteroids and comets can be affected by various forces, including the gravitational pull of planets, the Yarkovsky effect (a result of the asteroid's rotation causing uneven heating and radiation emission), and even collisions with other space objects. These effects can alter the orbits of NEOs over time, potentially increasing the risk of collision.

Tracking the orbits of NEOs is a continuous process, and advanced software and computer models are used to predict their paths with increasing accuracy. The more data collected on an object's orbit, the better scientists can estimate its future trajectory, including the possibility of an impact with Earth.

The identification and tracking of NEOs are not the efforts of individual nations or organizations but are part of a broader, international collaboration. Recognizing the global significance of NEO threats, various space agencies and organizations have come together to share data, coordinate efforts, and develop strategies for planetary defense.

NASA's Planetary Defense Coordination Office (PDCO) is a key player in the detection and tracking of NEOs. The PDCO works closely with observatories, researchers, and space missions to identify potential hazards and develop response plans.

The European Space Agency (ESA) also plays a crucial role in asteroid research and tracking through initiatives like the Near-Earth Object Program, which collaborates with both governmental and private entities to monitor NEOs.

The Minor Planet Center (MPC), operated by the International Astronomical Union (IAU), is a central hub for collecting and disseminating data on NEOs. It serves as a clearinghouse for observational data and provides valuable resources for astronomers worldwide.

These collaborative efforts have led to significant progress in identifying NEOs. As of recent years, astronomers have discovered thousands of NEOs, with new objects being catalogued regularly. The development of advanced computational models, improved telescopic surveys, and international partnerships has created a global network dedicated to planetary defense.

While significant progress has been made in identifying and tracking NEOs, challenges remain. Many smaller NEOs—those under 140 meters in diameter—are still difficult to detect, and a significant portion of the NEO population remains undiscovered. These smaller objects, though less likely to cause global catastrophes, could still cause significant regional damage if they were to collide with Earth.

The development of more sensitive telescopes, both ground-based and space-based, will be crucial in detecting smaller NEOs. Future missions, such as NASA's DART

(Double Asteroid Redirection Test) and ESA's HERA mission, aim to better understand the characteristics of NEOs and test methods of deflection and mitigation.

The identification and tracking of Near-Earth Objects are fundamental to understanding and mitigating the risks posed by asteroid impacts. As NEO detection methods improve, the ability to monitor and respond to these objects will enhance planetary defense capabilities, potentially preventing catastrophic collisions in the future. Through continued international cooperation, advanced technology, and sustained research efforts, humanity can better prepare for the challenges posed by these cosmic threats, ensuring the safety of our planet for future generations.

2.5. The Yarkovsky Effect and Orbital Drift

The Yarkovsky impact, a subtle however profoundly influential phenomenon in celestial mechanics, plays a crucial function in the long-term evolution of small our bodies in the sun device, particularly asteroids. Named after the Russian engineer Ivan Osipovich Yarkovsky, who first proposed the idea inside the early twentieth century, this effect describes how anisotropic thermal emissions—variations in how an object re-radiates warmness—can reason slow however substantial changes in the orbit of a rotating body in space. Although the pressure generated by way of the Yarkovsky impact is minuscule, over time it accumulates, leading to orbital flow that may determine whether or not an asteroid stays in a strong orbit or eventually crosses paths with Earth or every other planet.

The simple mechanism in the back of the Yarkovsky impact is rooted in thermodynamics and the interplay among sun radiation and a rotating frame. As sunlight strikes the floor of an asteroid, the floor heats up. Because maximum asteroids aren't perfectly reflective, a sizable portion of the incident solar strength is absorbed and later re-emitted as thermal infrared radiation. However, this re-radiation does now not occur uniformly throughout the surface or without delay after absorption. Due to the rotation of the body and the thermal inertia of the floor cloth, the most temperature—and therefore the height emission—happens not on the point of direct sun occurrence but slightly offset, typically in the afternoon area of the rotating item. This anisotropic thermal emission results in a small recoil pressure, just like a photon-driven thruster, which, relying on the route of rotation, can both increase or decrease the orbital semi-foremost axis.

The Yarkovsky impact is most tremendous for small asteroids, usually under 20 kilometers in diameter. Larger our bodies have extra mass and surface vicinity, diluting the have an impact on of thermal re-radiation on their momentum. The impact will become particularly vital in understanding the long-time period orbital evolution of Near-Earth Objects (NEOs), specifically those which are capacity impact hazards. Even a drift of some kilometers in keeping with yr can collect into a sizeable positional deviation over many years or centuries, appreciably altering impact possibilities and deflection techniques.

There are foremost components of the Yarkovsky impact: the diurnal and the seasonal consequences. The diurnal impact arises from the day by day rotation of the asteroid, analogous to Earth's day-night cycle. In this situation, the behind schedule re-radiation of heat causes a thrust in the course opposite to the afternoon aspect of the object. For a prograde-rotating asteroid (spinning inside the identical route as its orbit), this results in a internet thrust in the direction of orbital motion, inflicting the asteroid to slowly spiral outward—its semi-primary axis will increase. For retrograde rotators, the thrust works towards the orbital motion, causing inward float.

The seasonal Yarkovsky effect, on the other hand, is extra vast for bodies with slower rotation or with poles which are almost perpendicular to their orbital aircraft. It arises from the lag in floor heating due to the object's orbital function around the Sun. In this scenario, the hemisphere experiencing summer season (and for that reason more prolonged sun publicity) re-emits greater warmness and imparts a balk pressure that can either boost up or decelerate the orbit depending on the orientation of the rotation axis relative to the orbital movement.

The significance of the Yarkovsky force relies upon on several elements:

1. Size and Shape: Smaller gadgets are more vulnerable to the Yarkovsky effect due to the fact their surface location-to-mass ratio is better. Shape irregularities can also enlarge or decrease the anisotropic thermal emission.

2. Rotation Rate and Axis Orientation: Faster spinning bodies have a extra pronounced diurnal Yarkovsky effect. The tilt of the spin axis (obliquity) relative to the Sun affects whether or not the drift is inward or outward.

3. Surface Properties: Thermal conductivity, warmth capability, and surface roughness appreciably affect how warmness is absorbed and re-emitted. Dusty or regolith-blanketed surfaces have a tendency to retain heat longer, editing the timing and route of thermal emission.

4. Orbital Distance: The nearer the object is to the Sun, the more solar energy it receives, growing the electricity of the Yarkovsky pressure. This is why the effect is more applicable in the internal sun system.

5. Material Composition: Dense, metal our bodies have specific thermal behaviors than porous, rocky ones, altering their thermal inertia and emission traits.

Direct dimension of the Yarkovsky impact has become feasible way to excessive-precision astrometric tracking the use of radar and optical observations. Some asteroids, including (6489) Golevka, have exhibited clean orbital shifts on account of the Yarkovsky effect. The asteroid's trajectory, tracked over several years, discovered an unexplained float of approximately 15 kilometers in keeping with yr, a change inconsistent with gravitational perturbations by myself. This discrepancy changed into matched with predictions from thermal modeling, confirming the presence of Yarkovsky-prompted waft.

The discovery of the Yarkovsky effect as a actual, measurable phenomenon has profound implications for planetary defense. When projecting the future paths of probably unsafe asteroids (PHAs), even small uncertainties in thermal houses and spin states can result in massive trajectory deviations over many years. For example, asteroid (99942) Apophis, initially notion to have a mild hazard of impacting Earth in 2029, became carefully monitored with the Yarkovsky effect in thoughts. Adjustments to its expected route, informed by using thermal observations and modeling, have since eliminated the threat for that stumble upon but highlighted the necessity of accounting for non-gravitational forces.

Another crucial implication lies within the understanding of asteroid family dispersal. After a collisional breakup event inside the asteroid belt, fragments begin to float underneath the affect of the Yarkovsky impact. This reasons a fan-like distribution of orbital elements over time, developing observable V-fashioned patterns in plots of semi-primary axis versus size. By modeling this dispersal and evaluating it to modern asteroid family distributions, researchers can estimate the age of collisional occasions and higher apprehend the records of the asteroid belt.

The Yarkovsky impact additionally plays a pivotal role within the delivery of asteroids from the principle belt into Earth-crossing orbits. Dynamical pathways known as resonances—inclusive of the 3:1 suggest-motion resonance with Jupiter or the ν6 secular resonance—function escape hatches. When an asteroid's orbit slowly drifts into such a resonances due to the Yarkovsky effect, gravitational interactions can unexpectedly modify its eccentricity, injecting it into the internal sun gadget. In this way, a method driven by sunlight and thermal emission acts as the long fuse for doubtlessly catastrophic occasions on Earth.

Efforts to mitigate asteroid threats need to recall the Yarkovsky effect in planning and execution. For kinetic impactor missions or gravity tractors, even put up-deflection modeling need to comprise thermal effects, as any trade in spin country or floor residences should alter the subsequent Yarkovsky glide. The 2022 DART (Double Asteroid Redirection Test) task, which successfully impacted the moonlet Dimorphos, provided valuable records no longer simply on momentum transfer however on how the ensuing ejecta and floor changes might impact destiny thermal emission patterns.

Advanced modeling of the Yarkovsky effect carries complex floor thermophysics, along with the "beaming" effect as a result of floor roughness and cavities. These micro-geometric functions can redirect thermal photons in desired directions, adding an extra layer of complexity. Laboratory experiments the use of infrared sensors and simulated regolith have attempted to validate such models, supplying crucial statistics for refining orbit predictions.

Another extension of Yarkovsky-related forces is the YORP effect (Yarkovsky–O'Keefe–Radzievskii–Paddack), which describes how thermal emission and reflected daylight can regulate an asteroid's spin fee and axis orientation through the years. This spin evolution feeds returned into the Yarkovsky effect itself: because the spin modifications, so does the thermal emission sample, modifying the route and value of the orbital drift. In extreme cases, YORP-prompted spin-up can motive rubble-pile asteroids to undergo rotational fission, creating new binary systems or shedding mass.

The ability to model the Yarkovsky effect as it should be is contingent upon obtaining detailed observational information: radar shape fashions, thermal infrared spectra, spin axis determinations, and albedo measurements. Space missions like OSIRIS-REx, which studied and sampled asteroid Bennu, have substantially more desirable our expertise by offering close-up observations of thermal conduct, floor texture, and compositional heterogeneity. Bennu, which famous a great Yarkovsky flow of approximately 284 meters consistent with year, exemplifies how essential this effect is in planning sample return and planetary defense strategies.

In theoretical studies, researchers use N-body simulations augmented with non-gravitational pressure modules to propagate asteroid orbits forward and backward in time, incorporating uncertainties in thermal parameters and spin states. Monte Carlo strategies generate statistical ensembles of feasible future positions, quantifying impact probabilities and estimating how adjustments in surface circumstance—say, because of micrometeorite influences or outgassing—would possibly regulate the course.

The have a look at of the Yarkovsky effect additionally intersects with meteoritics. Many meteorites discovered on Earth are fragments of NEOs whose orbits had been slowly altered over tens of millions of years through thermal forces. By reading cosmic ray publicity a long time and orbital reconstructions, scientists can trace their journey from the asteroid belt, thru resonances, and ultimately into Earth's environment. This ties together the microphysical (radiative) and macro-orbital (trajectory) scales in a unified narrative.

In a broader astrophysical context, the standards underlying the Yarkovsky impact are not limited to our sun machine. Exoplanetary structures with dirt belts, minor planets, and particles disks may also exhibit similar thermal-balk-triggered migration of small our bodies. Understanding these forces should illuminate approaches of planetary device evolution and particles subject dynamics round other stars.

The Yarkovsky effect, once an obscure theoretical concept, has matured into a cornerstone of asteroid technology and planetary protection. It bridges the domain names of thermal physics, celestial mechanics, observational astronomy, and area challenge layout. Its implications ripple across time scales starting from decades to eons, and its affect stretches from the asteroid belt to Earth's skies. Whether modeling impact dangers or making plans asteroid aid missions, accounting for the Yarkovsky effect is essential. It reminds us that even the maximum delicate physical strategies, barely perceptible within the moment, can determine the fate of worlds.

2.6. The Role of Binary Asteroids in Collision Risk

Within the complex surroundings of our Solar System, small our bodies along with asteroids do now not all exist in isolation. A vast part of close to-Earth asteroids, specifically the ones large than 2 hundred meters, are simply binary systems—bodies gravitationally sure, orbiting every different as they together tour across the Sun. Understanding the function those binary asteroids play in the overall collision hazard with Earth is important. Their complex gravitational dance introduces specific demanding situations to predicting trajectories, assessing impact possibilities, and planning mitigation strategies, distinguishing them markedly from solitary asteroids.

Binary asteroids consist of a number one body, regularly substantially large, and a smaller secondary accomplice that orbits the primary at varying distances. Observational evidence amassed over the last few a long time—in particular thru radar imaging, light curve analysis, and recent spacecraft missions—has confirmed that about 10 to twenty percent of near-Earth asteroids belong to binary or multiple structures. These structures show off a number of configurations: a few have the secondary orbiting intently, tidally locked in order that the same face always factors towards the primary; others show more loosely bound mutual orbits, and a small subset even includes triple asteroid systems. This variety complicates efforts to version their lengthy-term dynamical conduct and their ability danger to Earth.

The formation of binary asteroids is typically connected to the YORP (Yarkovsky–O'Keefe–Radzievskii–Paddack) impact, a diffused but powerful pressure exerted by means of solar radiation on irregularly fashioned asteroids. Over time, choppy thermal emission due to daylight absorption and re-radiation can boom the asteroid's spin fee. When the rotation hastens sufficiently, centrifugal forces conquer the body's structural integrity, causing fabric to shed from its surface or main the asteroid to split into two gravitationally certain additives. This "rotational fission" technique is extensively generic because the dominant formation mechanism for binaries within the near-Earth population. Other formation scenarios include tidal disruptions throughout close planetary encounters, collisional fragmentation observed by gravitational reaccumulation, and primordial formation inside the early Solar System.

The dynamics within a binary device are inherently greater complicated than for single asteroids. The our bodies exert mutual gravitational forces, whilst both respond to the Sun's gravity and different perturbations from planets. Additionally, both the Yarkovsky and YORP effects affect the orbital evolution of the binary as a whole, as well as the mutual orbit among the additives. An mainly interesting phenomenon known as the Binary YORP (BYORP) impact arises from anisotropic thermal emission inside the binary device, that may reason the mutual orbit to enlarge, agreement, or maybe destabilize. This effect introduces an additional layer of complexity in predicting the destiny conduct of binaries.

Such multifaceted interactions render lengthy-term orbital predictions for binary asteroids hard. Whereas single asteroids already pose difficulties due to chaotic perturbations, thermal forces, and close planetary encounters, binary structures exacerbate those uncertainties. This heightened unpredictability complicates effect danger checks, which rely closely on specific trajectory forecasts.

One amazing aspect growing collision danger is the multiplied effective move-segment of binary systems. Having bodies orbiting in tandem essentially doubles—or more—their spatial quantity relative to a solitary asteroid of similar length. This more powerful vicinity increases the chance that one or each additives ought to intersect Earth's route. Moreover, mutual gravitational interactions can purpose the binary additives to separate or alternate orbits following near planetary encounters, potentially converting a unmarried binary system into multiple independent impactors.

During atmospheric entry, the presence of a binary machine leads to extra complex impact phenomena. Unlike a solitary asteroid that generally produces one primary airburst and effect web page, a binary may create multiple factors of fragmentation and explosion, relying at the relative sizes, velocities, and access angles of its components. This multiplicity complicates damage predictions and emergency reaction making plans, as multiple effect zones can spread destruction over a much wider region.

Detecting and characterizing binary asteroids is itself a disturbing observational venture. Light curve measurements frequently screen telltale signatures inclusive of mutual eclipses and occultations, however disentangling those alerts from noise can be difficult. Radar imaging presents definitive evidence of binary nature and permits direct measurements of sizes, shapes, and orbital parameters. However, such specific observations are resource-in depth and available best for a confined subset of recognized asteroids. Consequently, many binary systems remain unrecognized, adding an detail of wonder to collision hazard models.

The surface houses and rotational states of binary additives affect how thermal forces act upon them. For instance, the distribution of regolith, surface roughness, and fabric composition can modulate the value and direction of the Yarkovsky and YORP consequences. Since those results force tons of the long-time period orbital evolution, their variation between the primary and secondary bodies must be integrated into sensible models. Simple extrapolations from unmarried asteroid models prove inadequate for binaries, necessitating superior simulations that account for mutual shadows, thermal coupling, and variable spin states.

A high instance highlighting the importance of binary asteroid dynamics in collision risk is the Didymos system. Comprising a kind of 780-meter number one and a one hundred sixty-meter secondary, Didymos was decided on because the goal for NASA's DART challenge, the first test of kinetic impactor generation for asteroid deflection. The project's intention was to alter the orbit of the smaller associate around the primary, demonstrating humanity's functionality to persuade a probably unsafe binary asteroid's trajectory. Data amassed from DART and its comply with-up venture, ESA's Hera, offer unheard of insights into the physical and dynamical houses of binary asteroids and enhance our capacity to version impact situations and mitigation techniques.

Other well-studied systems, which includes (66391) 1999 KW4, function valuable herbal laboratories to understand binary asteroid evolution. Observations of these systems affirm theoretical predictions about spin-up strategies, BYORP-prompted orbital modifications, and long-term balance. Such studies underscore the necessity of inclusive of binary dynamics and thermal forces in collision danger analyses.

From a planetary defense attitude, the lifestyles of binary asteroids demands reconsideration of mitigation strategies. For instance, kinetic impactors have to be designed to have an effect on both additives effectively or at the least keep away from accidentally disrupting the binary structure, that can result in fragmenting the machine and growing the quantity of risky items. Accurate characterization of the binary's orbital parameters, mass distribution, and structural cohesion becomes essential to a success deflection missions.

Technological advances continue to beautify our information of binary asteroids. Improved radar structures, wide-discipline optical surveys, and infrared observations offer richer datasets for identifying and characterizing these systems. Coupled with sophisticated numerical fashions that simulate thermal forces, gravitational interactions, and material residences, scientists can higher predict how binary asteroids evolve and pose threats to Earth.

Future spacecraft missions are deliberate to analyze binary asteroids in extra detail, aiming to map their surfaces, degree internal structures, and reveal orbital adjustments over time. These efforts will refine our understanding of the way binaries shape, how they react to outside forces, and the way great to counter capability collision threats.

Binary asteroids play a vital role in the dynamics of close to-Earth item populations and pose specific demanding situations for effect chance evaluation. Their complicated gravitational interactions, more suitable effective go-sections, and sensitivity to thermal forces make predicting their trajectories greater uncertain than for solitary asteroids. Incorporating the physics of binary systems into planetary protection frameworks is fundamental for protective Earth from asteroid impacts. As observational abilties and theoretical models retain to enhance, our potential to understand and mitigate the dangers posed through those celestial couples will fortify, ensuring a more secure destiny for our planet.

CHAPTER 3

Effects and Traces of Asteroid Collisions

3.1. Gravitational Effects and Mass Extinctions

When an asteroid collides with a celestial frame, together with a planet or moon, a extraordinary quantity of energy is released, resulting in numerous unfavourable forces. The magnitude of the strength launch depends on the scale, velocity, and composition of the impacting asteroid. Here are the key aspects of the release of power during an effect:

Kinetic Energy: The primary source of strength at some stage in an impact is the kinetic power of the incoming asteroid. As the asteroid actions thru space at excessive velocities, this strength is transformed into detrimental forces upon effect.

Shockwaves: The kinetic electricity of the asteroid is unexpectedly transferred to the target frame, growing shockwaves that propagate via the impacted object. These shockwaves generate seismic waves, inflicting intense shaking and ground movement, just like earthquakes.

Crater Formation: The impact of the asteroid excavates material from both the impacting frame and the target body, forming a crater. The length and intensity of the crater depend upon the scale and strength of the impactor.

Ejecta: A large amount of cloth is ejected from the effect website at some stage in the collision. This ejecta includes each fragmented fabric from the impacting asteroid and particles from the goal frame.

Heat and Vaporization: The excessive strain and heat generated at some point of the impact can purpose the impacted substances to vaporize. This method releases monstrous amounts of energy and effects in the formation of a hot, sparkling plasma plume that rises from the impact site.

Fireball and Thermal Radiation: If the impactor is large enough, a fireball might also shape within the ecosystem due to the rapid compression and heating of the air in the front of the asteroid. The fireball emits intense thermal radiation.

Tsunamis: In the case of asteroid impacts in water our bodies, which includes oceans or seas, tsunamis can be generated. The effect strength displaces big amounts of water, ensuing in large waves that propagate outward from the impact web site.

Atmospheric Effects: Large asteroid impacts can inject massive amounts of dust and particles into the surroundings. This can cause a temporary cooling effect referred to as "impact winter," causing a drop in temperatures and disrupting ecosystems.

Understanding the release of power all through affects is crucial for assessing the capacity outcomes of asteroid collisions with Earth or different celestial bodies. While massive effect occasions have the capacity for catastrophic results, smaller influences also can cause localized harm and disruption. The study of effect events gives precious insights into the geological and environmental records of our planet and the wider solar machine.

When a big asteroid collides with an ocean or sea, it could generate effective waves called tsunamis. These tsunamis end result from the fantastic energy released all through the impact, which displaces water and sends waves propagating outward from the effect website online. Here are the key aspects of tsunamis and megatsunamis formation:

Tsunamis Formation:

Impact Energy: The impact of a big asteroid right into a body of water effects inside the speedy switch of electricity to the water. The mammoth force displaces a large extent of water from the impact website online.

Wave Propagation: The displacement of water creates a series of waves that radiate outward from the effect point. These waves travel at high speeds across the ocean or sea.

Amplification: As the tsunami waves method shallower coastal regions, their strength turns into compressed, inflicting the waves to grow taller and extra powerful.

Coastal Impact: Tsunamis can inundate coastal areas with huge waves, causing substantial flooding and devastating results for coastal groups and ecosystems.

MegaTsunamis Formation:

In uncommon instances, extremely massive impacts, together with those from pretty massive asteroids or comets, can create megatsunamis. These megatsunamis are a long way extra effective and negative than standard tsunamis because of the unheard of strength launch. Here are the traits of megatsunamis:

Enormous Impact: Megatsunamis are the end result of highly effective impacts, frequently with impactors many kilometers in diameter. The larger the impactor, the greater the energy released.

Global Consequences: Megatsunamis will have worldwide outcomes, sending waves throughout whole oceans and impacting coastlines worldwide. They can purpose large disruptions to marine and coastal environments.

Geological Signatures: Megatsunamis can depart distinctive geological signatures on coastlines, including deposits of sediment and debris some distance inland from the prevailing shoreline.

It's vital to be aware that whilst tsunamis and megatsunamis may be extraordinarily devastating, they are incredibly uncommon events. The probability of a big asteroid impact generating a megatsunami is low. Nonetheless, the observe of beyond impact activities and their results on Earth's geological file offers valuable insights into the capacity effect dangers and the want for ongoing efforts to screen and mitigate asteroid threats.

3.2. Impact Craters and Traces

The formation of craters on celestial our bodies, which include planets, moons, and asteroids, is the end result of various impact approaches. When an object, such as an asteroid or comet, collides with the floor of a strong body, it releases an immense amount of power. This energy reasons the excavation and change of the impacted surface, main to the formation of a crater. Here are the important thing mechanisms of crater formation:

Compression and Shockwaves: Upon impact, the impacting item swiftly decelerates and compresses the cloth on the factor of contact. This compression generates shockwaves that propagate via the impacted frame. The shockwaves create intense pressures and stresses, causing the target cloth to fracture and deform.

Excavation: A sizeable part of the effect electricity is used to excavate fabric from each the impacting object and the goal frame. Material from the impactor and the instantaneous surroundings of the effect website is ejected, forming a transient cavity.

Rebound: Following the excavation segment, the compressed and fractured material rebounds outward from the impact website online. This material may be thrown several kilometers faraway from the crater, specifically within the case of big affects.

Modification: The rebounded fabric reimpacts the encompassing terrain, modifying the crater's shape and inflicting secondary influences and structural deformation of the encompassing region.

Crater Collapse: For large craters, the hollow space created at some stage in excavation can also ultimately fall apart beneath the affect of gravity. The fall apart can result in the formation of a relevant peak or a vital uplift in the middle of the crater.

Ejecta Blanket: The cloth ejected during the impact settles back onto the surface surrounding the crater, forming an "ejecta blanket." The ejecta blanket includes a mixture of impactor and target cloth and can amplify over a extensive location.

Rim and Wall Formation: As the crater paperwork, the outer edges of the excavation disintegrate to shape the crater's rim and partitions. The length and peak of the rim depend on the impact power and the properties of the goal material.

Secondary Cratering: During the effect event, a few ejected material may also reimpact the floor, growing smaller secondary craters at a distance from the primary crater.

The mechanisms of crater formation play a critical role in shaping the landscapes of celestial our bodies. Craters offer precious facts about the history of effect activities and the geological processes that have formed a planetary frame's floor over the years. The look at of effect craters is critical for understanding the effect history of our sun system and the potential effect dangers faced by using Earth and other celestial our bodies.

Earth's geological history is marked by way of severa impact occasions, and the planet bears the scars of beyond asteroid and comet collisions. Many effect craters have been diagnosed and studied through scientists, providing treasured insights into the results of huge impacts on our planet's surface. Here are a few notable impact craters on Earth:

Chicxulub Crater (Mexico): The Chicxulub Crater is one of the most well-known impact craters on Earth and has had a big effect on our know-how of mass extinction activities. It is positioned off the coast of the Yucatán Peninsula in Mexico and is approximately a hundred and fifty kilometers (93 miles) in diameter. The effect that shaped this crater passed off around sixty six million years in the past and is broadly believed to be accountable for the mass extinction occasion that led to the death of the dinosaurs and plenty of different species.

Vredefort Crater (South Africa): The Vredefort Crater, positioned in South Africa, is one of the most important and oldest regarded impact systems on Earth. It has a diameter of about three hundred kilometers (186 miles) and is expected to be round 2 billion years vintage. The Vredefort effect occasion had a full-size geological effect and is related to the formation of gold deposits within the location.

Sudbury Basin (Canada): The Sudbury Basin in Ontario, Canada, is certainly one of the biggest impact craters on Earth, with a diameter of about one hundred thirty kilometers (81 miles). It is believed to have shaped approximately 1.85 billion years in the past. The impact that created the Sudbury Basin is associated with the Sudbury Igneous Complex, a wealthy source of nickel, copper, and other precious metals.

Popigai Crater (Russia): The Popigai Crater, placed in Siberia, Russia, is a fairly younger impact crater, envisioned to be approximately 35 million years old. It has a diameter of about a hundred kilometers (62 miles). The effect occasion at Popigai is assumed to have contributed to the formation of effect diamonds, additionally referred to as "Popigaiite."

Manicouagan Crater (Canada): The Manicouagan Crater in Quebec, Canada, is a prominent multiring effect crater with a diameter of about 100 kilometers (62 miles). It is expected to be approximately 214 million years vintage and is associated with an impact event that occurred at some point of the TriassicJurassic boundary.

Chesapeake Bay Crater (USA): The Chesapeake Bay Crater is an effect structure placed off the coast of Virginia, USA. It has a diameter of approximately 85 kilometers (fifty three miles) and turned into shaped approximately 35 million years in the past. The impact event created the most important regarded effect crater inside the United States.

These are only a few examples of the excellent impact craters on Earth. Impact events have played a full-size role in shaping our planet's geological records, and studying these craters offers valuable statistics about the dynamics of impact tactics and their outcomes on Earth's environment and lifestyles all through time.

3.3 Long-Term Environmental Changes and Climate Impact

Asteroid collisions, particularly large ones, are among the most powerful and potentially destructive events in the history of our planet. While the immediate impact of such an event can be catastrophic, the long-term environmental and climatic effects are equally profound, often reshaping the Earth's ecosystems for millions of years. The impact of an asteroid, especially one large enough to cause global consequences, can trigger a cascade of environmental changes that affect the planet's atmosphere, climate, and biodiversity in ways that are still being studied today.

When an asteroid collides with Earth, especially one of considerable size, the initial impact releases an enormous amount of energy. The kinetic energy from the asteroid is converted into heat, light, and mechanical energy, creating a range of immediate effects. These include:

Massive Wildfires: The intense heat from the impact can ignite vast wildfires, consuming large areas of forests and grasslands. These fires produce smoke and soot, which are injected into the atmosphere, reducing sunlight and affecting global temperatures.

Tsunamis and Shockwaves: For asteroids impacting the ocean, massive tsunamis can be triggered, causing widespread flooding and destruction along coastlines. The shockwaves from the impact also cause tremors and earthquakes, leading to further environmental disruption.

Ejection of Debris: The impact also sends vast amounts of debris, including vaporized rock and dust, into the atmosphere. Some of this debris will fall back to Earth as "impact ejecta," creating a blanket of material that can darken the skies and affect local climates.

These immediate effects, while devastating, are often just the beginning. The long-term environmental changes triggered by the collision can have far-reaching and enduring consequences on a global scale.

One of the most significant long-term effects of a major asteroid collision is the potential for global cooling, often referred to as a "nuclear winter." This scenario is based on the idea that the debris ejected by the impact would create a dense cloud of dust and soot that could obscure the Sun for months or even years. The blocking of sunlight would lead to a significant drop in global temperatures, causing widespread disruption to the Earth's climate.

The reduction in sunlight would severely impair photosynthesis, the process by which plants convert sunlight into energy. With less sunlight reaching the Earth's surface, plant life would struggle to survive. This collapse of plant life would have a domino effect on ecosystems, as herbivores would lose their food sources, followed by carnivores who depend on herbivores for sustenance. The collapse of food chains would devastate biodiversity and could lead to the extinction of numerous species.

The inability to grow crops would also lead to food shortages, impacting human populations and animal life. In this scenario, the Earth could experience a prolonged agricultural crisis, which might last years or even decades, depending on the size of the asteroid and the extent of the climatic disruption.

Global temperatures would likely plummet due to the ash, soot, and vaporized rock particles that would be ejected into the atmosphere. These particles would block incoming solar radiation, leading to a dramatic cooling of the Earth's surface. In some models, temperatures could drop by as much as 10-20°C (18-36°F), significantly altering global climate patterns.

This cooling effect could resemble the conditions seen during the "Little Ice Age," a period of colder-than-usual temperatures in Europe and North America that lasted from the 14th to the 19th centuries. However, the scale and intensity of the cooling caused by an asteroid impact would be much greater and more widespread, affecting global weather patterns, including the potential for severe winters and reduced precipitation in some areas.

Another significant consequence of asteroid impacts is the release of sulfur and carbon dioxide into the atmosphere. The intense heat generated by the impact can vaporize sulfur-rich minerals, creating sulfur dioxide gas. When sulfur dioxide combines with water vapor in the atmosphere, it forms sulfuric acid, which falls as acid rain.

Acid rain can have severe environmental effects, including the destruction of plant life, the contamination of freshwater sources, and the disruption of aquatic ecosystems. Additionally, the increase in atmospheric carbon dioxide levels from the impact could lead to long-term global warming after the initial cooling period. Carbon dioxide is a potent greenhouse gas, and its accumulation in the atmosphere could trap heat, leading to a gradual increase in global temperatures.

The combination of cooling and eventual warming could create significant instability in the Earth's climate system, with unpredictable weather patterns and more frequent extreme weather events.

The long-term environmental changes caused by asteroid impacts are also closely linked to mass extinctions. The most famous example of this is the mass extinction event that occurred approximately 66 million years ago, marking the end of the Cretaceous period and the extinction of the dinosaurs. This event is widely believed to have been caused by the impact of a large asteroid, which led to a dramatic transformation of the Earth's climate and ecosystems.

The asteroid impact that contributed to the extinction of the dinosaurs is thought to have caused both immediate and long-term environmental changes. The initial impact likely caused widespread fires, earthquakes, and tsunamis, while the long-term effects, such as the "nuclear winter" scenario, led to a drastic cooling of the planet. This cooling, combined with the reduced availability of sunlight, disrupted the food chains and ecosystems that supported the dinosaurs.

In addition to the cooling, the release of sulfur and other gases into the atmosphere likely led to acid rain, further devastating plant and animal life. The combination of these factors led to the extinction of nearly 75% of all species on Earth, including the non-avian dinosaurs. This event marked one of the most significant mass extinctions in Earth's history, reshaping life on the planet and allowing mammals to become the dominant group of animals in the subsequent era.

While the extinction of the dinosaurs is the most well-known result of an asteroid collision, other mass extinctions in Earth's history may have also been influenced by asteroid impacts. Evidence suggests that several large scale extinctions occurred during the Late Ordovician, Late Devonian, and

Permian periods, some of which may have been triggered or exacerbated by asteroid collisions.

After each mass extinction, life on Earth took millions of years to recover. The extinction events created ecological opportunities for new species to evolve and fill the vacant ecological niches. In the case of the Cretaceous-Paleogene extinction event, the rise of mammals, including primates and eventually humans, was made possible by the disappearance of dinosaurs and other dominant species.

Given the catastrophic potential of asteroid impacts, scientists have developed models to predict the long-term environmental effects of future asteroid collisions. These models simulate the impact process, including the amount of energy released, the distribution of ejecta, the potential for global cooling, and the subsequent warming from greenhouse gases.

By understanding the likely effects of asteroid impacts, scientists can develop more effective strategies for impact mitigation, early warning systems, and post-impact recovery. These models also provide valuable insight into the Earth's past, helping researchers to understand how previous impact events have shaped life on our planet.

The long-term environmental changes and climate impacts resulting from asteroid collisions can reshape the Earth in profound ways. From global cooling and acid rain to mass extinctions and long-term recovery, the effects of an asteroid collision can persist for millions of years. The understanding of these impacts is not only critical for planetary defense but also for gaining insight into the Earth's history and the dynamic processes that have shaped life on our planet. While asteroid collisions are rare, their potential for global destruction underscores the importance of ongoing research, monitoring, and preparation for such events.

3.4. Psychological Impact of Global Threats

Global threats, ranging from climate exchange and pandemics to nuclear proliferation and huge-scale natural disasters, pose immense challenges now not only to bodily protection and survival however also to the collective and individual psyche of humanity. The awareness of such looming risks has profound mental ramifications, influencing mental health, societal behavior, and cultural dynamics. Understanding the psychological effect of world threats is essential for growing powerful conversation strategies, enhancing resilience, and designing interventions that aid individuals and communities for the duration of intervals of heightened tension and uncertainty.

The human mind is stressed to respond to threats as part of an evolutionary survival mechanism. Acute threats cause a cascade of physiological and psychological responses— regularly described as the "combat, flight, or freeze" reaction—

that prepare the organism for fast motion. However, international threats inclusive of weather exchange or the chance of nuclear struggle range essentially from on the spot risks. They are frequently characterised by means of their continual, complex, and intangible nature, unfolding over lengthy durations and affecting complete populations. This temporal and spatial scale creates a mental paradox: whilst the hazard is real and enormous, it is able to experience abstract or remote, main to unique forms of pressure and coping responses.

One essential mental effect of world threats is the rise of continual stress and anxiety. Unlike acute pressure, which is severe however brief-lived, chronic pressure related to continual threats can erode mental properly-being over the years. Individuals may revel in heightened emotions of helplessness, worry, and uncertainty approximately the future. This sustained tension can appear in signs consisting of insomnia, concentration difficulties, irritability, and somatic complaints. In excessive instances, extended exposure to international danger cognizance can make a contribution to melancholy and other mental health disorders.

The concept of "eco-anxiety" has won prominence in recent years as a selected shape of anxiety related to environmental and climate-related worries. Particularly among younger generations, who face the possibility of living in a world increasingly shaped by using environmental degradation, this anxiety is related to emotions of grief, depression, and existential dread. Surveys screen that significant quantities of adolescents populations report worry approximately the future of the planet, that may from time to time result in disengagement or fatalism if not addressed constructively.

Another measurement of the mental effect relates to perceived threat and its have an effect on on conduct. Humans do not always reply rationally to statistical possibilities or medical projections. Cognitive biases inclusive of optimism bias, denial, and affirmation bias often shape how human beings perceive and react to threats. For example, individuals might downplay the severity of climate change as it conflicts with their financial hobbies or political affairs. Such mental defenses can serve adaptive purposes by means of decreasing anxiety however can also prevent collective motion and coverage guide.

Mass media and social networks play a pivotal function in shaping public perceptions of worldwide threats. Sensationalized reporting can amplify fear and panic, whilst misinformation and conspiracy theories can breed confusion and mistrust. Conversely, clean and constant communique that emphasizes organization, preparedness, and solutions can empower people and foster resilience. The psychological framing of messages—whether or not specializing in doom and catastrophe or on hope and motion—extensively affects emotional responses and willingness to engage in protecting behaviors.

The social material itself may be strained underneath the load of world threats. Collective tension can cause social polarization, scapegoating, and warfare, as communities search for causes and assign blame. Conversely, shared demanding situations also have the capacity to foster team spirit, cooperation, and collective resilience. Historical examples from wartime mobilization to pandemic responses illustrate the potential of societies to unite around common desires within the face of existential chance.

Individual differences in psychological resilience play a essential role in determining how human beings deal with global threats. Factors such as personality tendencies, social help networks, earlier trauma, and cultural context affect vulnerability and adaptability. Psychological resilience isn't truely an innate trait but can be cultivated via interventions consisting of strain control schooling, community engagement, and fostering a feel of motive and control.

The place of work and educational environments are critical arenas wherein the mental impacts of world threats are both felt and addressed. For example, educators incorporating climate literacy into curricula ought to stability informing students approximately dangers with supplying equipment for optimistic movement to avoid overwhelming them. Similarly, employers and policymakers want to take into account mental health help structures as a part of disaster preparedness and response planning.

Global threats also venture philosophical and existential frameworks that individuals rely upon for that means and safety. Questions about humanity's vicinity inside the international, the fairness of herbal activities, and the opportunity of a sustainable future come to the fore. These existential anxieties can show up as spiritual crises or renewed interest in meaning-making practices, including faith, philosophy, or activism.

From a scientific psychology attitude, addressing the intellectual fitness outcomes of global threats requires a multidisciplinary method. Interventions ought to be sensitive to the specific capabilities of persistent, large-scale stressors, incorporating network-based totally packages, trauma-knowledgeable care, and rules that address social determinants of health. Prevention strategies emphasizing early education, public focus, and empowerment can mitigate the onset of excessive mental distress.

The position of era in modulating mental responses to global threats is a double-edged sword. On one hand, virtual structures facilitate information dissemination, social help, and mobilization of collective motion. On the other hand, they can contribute to "doomscrolling," records overload, and the unfold of misinformation. Digital literacy and vital questioning competencies turn out to be vital tools in navigating the complex emotional landscape fashioned by worldwide threats.

In recent a long time, studies into post-traumatic boom presents hopeful insights into how people and communities can't only survive but also find new electricity and that means after confronting adversity. Global threats, whilst overwhelming, additionally offer possibilities for transformation. Cultivating narratives of hope, emphasizing human ingenuity and harmony, and fostering adaptive coping techniques are key to harnessing this capability.

The mental effect of world threats therefore includes a large spectrum of responses—from anxiety and depression to resilience and boom. It demanding situations individuals, communities, and societies to rethink priorities, values, and strategies for coexistence. Understanding those psychological dimensions is essential to constructing a world which can successfully navigate gift and future threats with both vigilance and wish.

3.5. Cultural Reflections of Historical Impact Events

Throughout human records, celestial impacts—asteroids, comets, and meteorites putting Earth—have not most effective fashioned the physical world but have also profoundly motivated human subculture. The dramatic and often catastrophic nature of those activities has permeated mythology, art, literature, faith, and collective memory, leaving indelible marks on how societies understand their area within the cosmos and the fragility of existence. The cultural reflections of ancient impact events monitor humanity's try to grapple with forces a long way past human manipulate, interpret existential threats, and are looking for meaning amid chaos.

The cultural imprint of impact events may be traced back hundreds of years, often intertwined with historical oral traditions and early written records. Many indigenous cultures recount testimonies of "hearth from the sky," "thunderstones," or "sky gods" wielding unfavorable power—interpretations that modern scholars now on occasion link to real impact phenomena witnessed with the aid of early peoples. These narratives served more than one functions: explaining surprising environmental modifications, caution towards moral transgressions, and reinforcing social concord through shared ideals.

One outstanding instance is the Tunguska occasion of 1908 in Siberia, which flattened an enormous woodland place. Though scientifically understood best within the current technology, it has come to be embedded in neighborhood folklore, inspiring myths of divine punishment and supernatural forces. The event additionally sparked global medical hobby and cultural curiosity, symbolizing the unpredictable strength of the cosmos.

Religious interpretations of effect occasions often framed them as acts of divine wrath or cosmic judgement. In numerous historic texts, tremendous floods, fiery destructions, and celestial upheavals resemble the aftermath of catastrophic affects. The biblical tale of the Great Flood, for instance, has been hypothesized by means of a few researchers as a mythologized reminiscence of an effect-induced tsunami or environmental catastrophe. Similarly, the destruction of towns in diverse mythologies—which includes Sodom and Gomorrah—has been speculated to mirror ancient debts of airbursts or meteor influences.

Artistic expressions offer a rich tapestry via which effect events had been imagined and memorialized. From prehistoric petroglyphs that a few students interpret as depictions of celestial phenomena to Renaissance paintings symbolizing apocalyptic destruction, art serves as a medium for processing fear, awe, and the elegant nature of cosmic forces. The motif of hearth raining from the heavens recurs throughout cultures, encapsulating humanity's simultaneous fascination and terror.

In literature, effect occasions regularly appear as metaphors for transformation, apocalypse, or rebirth. The topic of cosmic disaster resonates with broader human issues approximately mortality, hubris, and renewal. Modern technology fiction often explores the results of asteroid affects, projecting modern anxieties and hopes regarding planetary defense and survival. Such narratives shape public notion and make contributions to the cultural talk approximately humanity's vulnerability and resilience.

Beyond myths and testimonies, effect activities have prompted ritual and lifestyle. Some cultures advanced ceremonies or taboos related to celestial phenomena, reflecting attempts to appease perceived supernatural dealers or to mark cycles of destruction and renewal. These practices strengthened social order and provided mental coping mechanisms within the face of uncontrollable natural disasters.

The study of cultural reflections additionally intersects with archaeology and anthropology, revealing how effect activities would possibly have triggered societal collapses or migrations. For example, the end of the Late Bronze Age coincides with evidence of sizable upheaval that a few students associate with weather disruptions doubtlessly because of extraterrestrial influences. While direct causation is debated, the interaction between effect occasions and human records remains a fertile area of inquiry.

The upward push of current scientific information transformed cultural narratives about effect activities. As astronomy and geology unveiled the mechanisms and frequencies of affects, the belief shifted from divine wrath to herbal chance. This medical framing brought new cultural meanings, emphasizing prevention, preparedness, and international cooperation. Public engagement with asteroid effect dangers often blends clinical literacy with lingering mythic elements, reflecting the deep-rooted vicinity of such occasions in human imagination.

In present day way of life, effect occasions inspire now not only worry however also fascination and hope. They function catalysts for global collaboration, inclusive of efforts to music near-Earth gadgets and increase deflection technology. Cultural products—movies, documentaries, books—train and entertain, elevating recognition whilst exploring profound philosophical questions on humanity's future.

Moreover, the mental resonance of impact events maintains to influence cultural expressions. Apocalyptic and submit-apocalyptic genres thrive in famous media, reflecting collective anxieties about existential threats. These narratives allow societies to discover situations of fall apart and renewal, to confront mortality, and to assume new beginnings.

The cultural reflections of historic impact activities span millennia and myriad paperwork, revealing the deep interconnection between humanity and the cosmos. They seize the awe-inspiring energy of celestial phenomena and the human quest to discover that means inside the face of cosmic unpredictability. Understanding those cultural dimensions enriches our appreciation of impact activities—no longer merely as clinical occurrences however as profound agents shaping human thought, identity, and creativity.

3.6. Oceanic Impacts and Tsunami Models

When celestial our bodies which includes asteroids or comets collide with Earth's extensive oceans, the effects are regularly as dramatic and far-achieving as land impacts, but they unfold with unique dynamics governed with the aid of the interactions among the impactor, the water column, and the coastal environments. Oceanic affects, although much less visible than terrestrial craters, have the capacity to generate large tsunamis, regulate climate styles, and disrupt marine and human ecosystems on scales that mission our predictive and response capacities.

Understanding oceanic affects calls for an interdisciplinary method, combining planetary technological know-how, oceanography, fluid dynamics, and threat modeling. The observe of such impacts has grown in urgency given the growing focus of near-Earth gadgets (NEOs) that might probably strike the oceans, which cowl over 70% of the Earth's surface. While direct remark of historic oceanic impacts is restricted, proof from geological deposits, tsunami sediments, and numerical simulations gives treasured insights into their mechanisms and outcomes.

The physics of an oceanic impact is complicated and markedly special from land influences. When an asteroid strikes the sea surface at high velocity—frequently tens of kilometers in step with 2nd—it generates an massive switch of kinetic strength to the water, generating a swiftly increasing vapor plume, shock waves, and a towering preliminary hollow space known as a temporary crater. This hollow space quickly collapses, riding upward and outward water flows that could spawn tsunamis with wave heights ranging from meters to hundreds of meters depending on impactor size, speed, and perspective.

Tsunami generation from impacts is encouraged with the aid of several key elements: the scale and density of the impactor, the depth of the sea on the impact site, and the bathymetry of the encircling seafloor. Large, dense impactors putting shallow waters generally tend to supply greater sizable waves through displacing greater volumes of water. The angle of effect additionally plays a important position—indirect impacts can create uneven wave patterns, focusing energy specifically guidelines and consequently affecting particular coastlines disproportionately.

Numerical fashions have end up critical equipment for simulating oceanic affects and ensuing tsunamis. Early models used simplified assumptions, treating water as an incompressible fluid and approximating effect dynamics with scaling laws derived from laboratory experiments and land impact analogs. Modern computational fluid dynamics (CFD) simulations contain state-of-the-art equations of nation for water and vapor phases, shock physics, and 3-dimensional bathymetric records to seize the temporary evolution of impact-generated waves with high fidelity.

One challenge in tsunami modeling is capturing the transition from deep ocean propagation to coastal inundation. While tsunamis in the open ocean frequently have small wave heights and long wavelengths, their energy concentrates dramatically as they method shallow coastal zones, causing speedy increases in wave amplitude and damaging capacity. Accurate fashions must consequently combine impact dynamics with hydrodynamic tsunami propagation and run-up calculations, accounting for coastal topography, tidal consequences, and man-made structures.

Historical geological evidence shows that oceanic affects have produced some of the maximum devastating tsunamis in Earth's records. Sediment cores from coastal and marine deposits screen layers regular with fast, excessive-strength water flows function of tsunami activities. For example, the Chicxulub effect 66 million years ago, widely related to the mass extinction of dinosaurs, is believed to have generated enormous tsunamis with waves exceeding hundreds of meters, affecting coastlines around the globe.

Modern tsunami danger exams now automatically encompass scenarios related to oceanic impacts. These checks tell disaster preparedness plans, early warning systems, and public schooling campaigns, specifically for vulnerable coastal populations. However, large uncertainties stay due to the rarity of massive affects and the complexities of oceanic wave dynamics, making probabilistic danger estimation a crucial however challenging endeavor.

International cooperation performs a essential position in monitoring potential oceanic impact threats. Organizations just like the International Asteroid Warning Network (IAWN) and the United Nations Office for Outer Space Affairs (UNOOSA) coordinate commentary efforts, records sharing, and contingency planning to mitigate risks posed via NEOs. Tsunami caution facilities international integrate seismic, oceanographic, and astronomical information to stumble on tsunami-generating events rapidly and difficulty alerts.

In addition to threat mitigation, the examine of oceanic impacts contributes to broader clinical questions on Earth's geological and biological evolution. The interplay among effect-generated tsunamis and marine ecosystems gives insights into beyond extinction activities, sediment delivery techniques, and the resilience of coastal habitats. Understanding those dynamics also informs the search for effect signatures in the geological record, aiding in the reconstruction of Earth's effect records.

Emerging research explores novel modeling strategies, inclusive of coupling excessive-resolution impact simulations with weather models to assess the lengthy-term atmospheric and oceanic consequences of oceanic influences. These included fashions observe how effect-precipitated tsunamis may additionally trigger remarks loops affecting ocean stream, carbon cycles, and global temperatures, with implications for information both beyond and capability future events.

Public interest in oceanic affects and tsunamis is frequently heightened following natural disasters inclusive of the 2004 Indian Ocean tsunami or the 2011 Tōhoku tsunami in Japan. These events underscore the devastating strength of large waves and the importance of preparedness, making the scientific look at of effect-generated tsunamis each well timed and socially relevant. Educational outreach and interdisciplinary research keep to extend our knowledge, decorate resilience, and inspire collaborative efforts to safeguard humanity from these profound cosmic dangers.

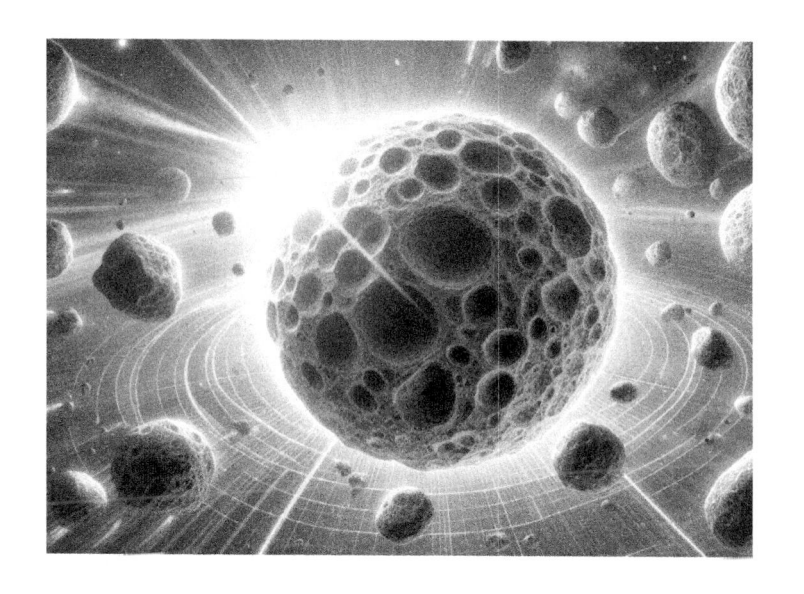

CHAPTER 4

Potential Collision Scenarios

4.1. Study of HighRisk Asteroids

As an AI language model, I don't have realtime browsing abilties, so I can't provide the maximum contemporary list of Potentially Hazardous Asteroids (PHAs). The listing of PHAs is often updated as new asteroids are determined and their orbits are better characterized.

To get right of entry to the state-of-the-art and maximum correct data approximately Potentially Hazardous Asteroids, I propose journeying professional web sites and databases maintained by using area agencies and astronomy corporations, together with NASA's NearEarth Object Program (CNEOS), the European Space Agency's NearEarth Object Coordination Centre (NEOCC), or the International Astronomical Union's Minor Planet Center (MPC).

These organizations constantly screen and track nearEarth objects, including PHAs, and offer uptodate data on their orbits, sizes, and ability effect dangers. By referring to their official web sites, you could find the most modern and dependable records approximately Potentially Hazardous Asteroids and their characteristics.

Impact opportunity and danger assessment are essential components of analyzing Potentially Hazardous Asteroids (PHAs) to recognize the potential dangers they pose to Earth and other celestial bodies. Scientists use numerous strategies and gear to evaluate the likelihood of an asteroid impact and compare the ability results if such an effect have been to occur. Here are the important thing components of effect probability and threat evaluation:

Impact Probability: Determining the opportunity of an asteroid effect involves calculating the probabilities of a specific PHA colliding with Earth within a given time-frame. Scientists use data on the orbits, sizes, and velocities of PHAs to are expecting their future trajectories and potential near strategies to our planet.

Orbital Uncertainty: The orbital predictions for PHAs have some stage of uncertainty because of factors along with incomplete observational statistics or the influence of gravitational interactions with planets. As a result, effect chance calculations regularly contain considering more than a few viable orbits (impact corridors) in place of a unmarried trajectory.

Torino Scale: The Torino Scale is a threat evaluation device used to talk the effect danger stage associated with a particular PHA. It ranges from zero to 10, with 0 indicating no danger and 10 representing a positive effect with devastating results. The Torino Scale helps officers and the general public recognize the potential threat posed by way of a selected PHA.

Palermo Technical Impact Hazard Scale: The Palermo Scale is another effect threat evaluation device that quantifies the general chance posed through a PHA. It takes under consideration the effect chance and potential energy released upon impact. A higher Palermo Scale value indicates a extra capability threat.

Risk Mitigation Strategies: Once the effect probability and risk related to a PHA had been assessed, scientists and policymakers can don't forget various mitigation strategies to divert or mitigate the potential effect. These techniques may also include gravitational tractor missions, kinetic impactors, or different strategies geared toward changing the asteroid's trajectory.

Frequency of Impact Events: By studying beyond effect events and the geological document, scientists can estimate the frequency of asteroid impacts on Earth. Understanding the frequency of effect occasions is essential for assessing the longterm impact hazard posed by means of PHAs.

Planetary Defense Efforts: Global efforts in planetary protection involve non-stop monitoring and tracking of nearEarth items, along with PHAs. Space groups and observatories international collaborate to improve our capability to stumble on, represent, and predict the orbits of PHAs to enhance our preparedness for potential effect threats.

Assessing the impact probability and threat of PHAs is an ongoing method that involves combining observational statistics, computational models, and threat assessment scales. By reading these factors, scientists can higher apprehend the ability impact risks and make informed decisions regarding planetary defense and hazard mitigation techniques.

4.2. Possible Collision Points and Potential Effects

Calculations and simulations play a essential role in the observe of Potentially Hazardous Asteroids (PHAs) and their capacity impact on Earth. Scientists use advanced computational strategies to perform various calculations and simulations, enabling them to better apprehend the orbits of PHAs, examine effect probabilities, and simulate effect situations. Here are the important thing aspects of calculations and simulations associated with PHAs:

Orbit Determination: Calculating the orbits of PHAs requires analyzing observational facts obtained from telescopes and radar observations. Scientists use numerical techniques and algorithms to suit orbital parameters to the observational records, permitting them to expect the future positions and trajectories of PHAs appropriately.

Impact Probability: To check the probability of a PHA colliding with Earth, scientists use statistical analyses and Monte Carlo simulations. Monte Carlo simulations contain randomly sampling from ranges of unsure parameters, together with orbital elements and observational mistakes, to create lots of feasible impact situations. By going for walks these simulations, scientists can estimate the likelihood of impact occasions within precise timeframes.

Impact Energy and Consequences: Simulating effect occasions allows scientists to calculate the energy released upon effect and investigate the capacity effects on Earth. The effect strength is associated with the asteroid's length, velocity, and composition. Simulations additionally assist version the formation of effect craters, the technology of tsunamis (if impacting an ocean), and the outcomes on the surroundings and environment.

Risk Assessment: Risk assessment includes integrating impact probability with impact results. Scientists use computational models to combine the possibility of an effect with the capacity damage because of the effect to quantify the general risk posed with the aid of a PHA. The effects are regularly expressed the use of scales like the Torino Scale or the Palermo Scale, as referred to in advance.

Trajectory Analysis: Numerical simulations of asteroid trajectories help expect the destiny motions of PHAs and assess their capacity near strategies to Earth. By studying these trajectories, scientists can perceive potential impact dangers, even for activities loads or lots of years into the future.

Impact Mitigation Strategies: Computational simulations are crucial for studying and evaluating impact mitigation strategies. Scientists use simulations to version the outcomes of diverse deflection strategies, which include deploying a spacecraft to regulate the asteroid's trajectory thru gravity assists or kinetic impactors. These simulations help decide the first-rate method to mitigate ability impact dangers.

Planetary Defense Preparedness: By using calculations and simulations, researchers can check the effectiveness of various detection and tracking techniques, enhancing our planetary protection preparedness. Regular simulations also assist refine response plans and emergency protocols in the unlikely occasion of a confirmed PHA effect danger.

In summary, calculations and simulations are fundamental gear for information PHAs, comparing their effect risks, and formulating appropriate planetary protection measures. These computational methods provide valuable insights into potential effect eventualities, allowing scientists and policymakers to make informed choices to guard our planet from ability asteroid threats.

When a Potentially Hazardous Asteroid (PHA) collides with Earth or every other celestial frame, the impact can trigger a chain of secondary results which have wideranging results. These secondary consequences stand up as a result of the big electricity released at some point of the effect occasion. Here are a number of the important thing secondary results that could occur after a collision:

Tsunamis: If the impact happens in an ocean or sea, it could generate tsunamis (huge ocean waves) that propagate throughout substantial distances. These tsunamis can inundate coastal areas, causing huge flooding and devastation.

Seismic Activity: The impact's shockwaves can create extreme seismic hobby, much like earthquakes. The electricity launched at some stage in the collision can purpose the floor to shake and cause floor ruptures inside the location of the impact website.

Crater Formation: The impact creates a crater on the surface of the impacted celestial body. The length and intensity of the crater depend on the dimensions and energy of the impacting object. For large affects, imperative peaks or vital uplifts might also form within the crater due to the rebound of compressed cloth.

Ejecta and Impact Debris: Material ejected from the effect web page at some stage in the collision can be scattered over large distances. The ejected material, known as impact particles or ejecta, can rain down on surrounding regions, contributing to secondary impacts.

Fire and Thermal Radiation: A massive impact can generate excessive heat and hearth at the impact website and its surroundings. The warmth generated for the duration of the effect can reason fires and burn plants inside the vicinity.

Atmospheric Effects: The effect event can inject a sizeable amount of dirt, soot, and gases into the surroundings. This can lead to shortterm atmospheric changes, including an "impact winter" effect, in which sunlight is blocked, ensuing in a cooling of the planet's floor.

Global Environmental Changes: Depending at the effect's size, the secondary results can motive significant worldwide environmental adjustments. Large influences have the ability to alter weather patterns, disrupt ecosystems, and contribute to mass extinctions.

MegaTsunamis: In the case of fairly massive affects, megatsunamis can be generated. These are a long way extra powerful than usual tsunamis and may have global outcomes.

It is critical to be aware that the value and extent of secondary consequences rely on the size, electricity, and area of the impact. While a few influences may purpose only localized outcomes, others may have global implications, especially for big PHAs. Studying beyond impact events and their consequences on Earth and different celestial our bodies gives precious insights into the ability secondary consequences of asteroid collisions and their impacts on the surroundings and life.

4.3 Modeling Impact Scenarios and Probabilities

Asteroid collisions with Earth represent one of the most significant threats in terms of their potential for widespread destruction. While the probability of such a catastrophic event occurring within a human lifetime is relatively low, the consequences of a large impact could be disastrous, affecting life on Earth in profound and far-reaching ways.

Modeling impact scenarios and calculating probabilities of asteroid collisions are essential for understanding the scale of potential threats and preparing for possible outcomes. Given the complexity and uncertainty surrounding asteroid orbits, impact effects, and the various environmental changes that could follow an impact, detailed models provide insights into both the likelihood of an impact and the potential consequences for Earth. These models also allow scientists and space agencies to design and evaluate mitigation strategies, improving preparedness for such events.

The primary challenges in modeling impact scenarios stem from the need to account for numerous factors, including the asteroid's size, speed, composition, and trajectory, as well as the location of the impact. Despite these challenges, advancements in computational modeling, observational technology, and impact research have significantly improved our ability to assess these risks and develop more accurate predictions.

Several factors need to be considered when modeling asteroid impact scenarios. These include the physical characteristics of the asteroid, the impact dynamics, and the resulting environmental consequences. The following are the key elements involved in impact modeling:

1. Asteroid Size and Composition: The size of the asteroid is one of the most critical factors in determining the severity of the impact. Larger asteroids carry more kinetic

energy and release more heat upon impact. However, the composition of the asteroid is equally important. Metallic asteroids, for example, are denser and more likely to survive the atmosphere than stony asteroids, which are more likely to disintegrate during entry.

2. Impact Velocity: The speed at which the asteroid strikes Earth is another crucial variable. Asteroids can travel at velocities ranging from 10 to 70 km/s, and the higher the velocity, the more energy is released upon impact. This energy is responsible for the generation of shockwaves, heat, and the formation of craters, all of which contribute to the severity of the impact.

3. Impact Angle and Location: The angle at which the asteroid strikes the Earth also plays a significant role in determining the consequences of the collision. An impact at a steep angle tends to produce more localized effects, while a shallow impact angle can lead to widespread devastation over a larger area. The location of the impact—whether it occurs in the ocean or on land—also influences the type of damage caused. Ocean impacts, for example, are more likely to generate massive tsunamis, while land impacts may create large impact craters and ejecta.

4. Atmospheric Effects: As the asteroid enters the Earth's atmosphere, friction generates intense heat, causing it to vaporize and release energy. The interaction between the asteroid and the atmosphere can also create shockwaves,

producing damage in the surrounding region. Additionally, the impact can eject debris into the atmosphere, which can obscure sunlight and lead to cooling effects, contributing to longer-term climatic changes.

5. Post-Impact Environmental Effects: After the immediate impact, the Earth would experience a range of environmental changes, including tsunamis, wildfires, atmospheric disruption, and long-term cooling or warming. These secondary effects can cause widespread environmental damage and potentially lead to extinction-level events, depending on the size and nature of the asteroid.

While asteroid impacts are relatively rare, the probability of such events occurring depends on several factors, including the size of the asteroid and the timeframe under consideration. There are various statistical models and methods used to estimate the likelihood of an impact, often based on observations of near-Earth objects (NEOs) and their trajectories.

Scientists use statistical models to estimate the frequency of asteroid impacts on Earth. These models take into account the number of asteroids in the solar system, the frequency with which they cross Earth's orbit, and the probability of collision. Generally, smaller asteroids (less than 140 meters in diameter) are more common and have a higher probability of entering Earth's atmosphere, while larger asteroids (over 1 km in diameter) are much rarer but pose a significantly greater threat.

The estimated probability of a large asteroid (greater than 1 km in diameter) impacting Earth is roughly 1 in 500,000 per year. However, for smaller asteroids, the frequency of impacts is higher. For example, an asteroid 140 meters in diameter or larger might strike Earth once every 20,000 years. While these events are statistically unlikely to happen during a human lifetime, the potential consequences of such an event justify continued research and monitoring.

To quantify the risk associated with asteroid impacts, scientists have developed various hazard scales. The Torino Scale is one such scale that categorizes the level of risk posed by a potential asteroid collision. It ranges from 0 (no risk) to 10 (a certain catastrophic impact). For example, a Torino Scale rating of 1 indicates a "slightly hazardous" impact, while a rating of 8 or higher would suggest an impact with the potential for global catastrophe.

The Palermo Technical Impact Hazard Scale is another tool used to assess the probability of an asteroid impact. This scale takes into account both the size of the asteroid and the probability of collision, providing a more detailed risk assessment. The Palermo scale is logarithmic, meaning that a score of -2 represents a very low likelihood of impact, while a score above 0 indicates an increased risk.

These hazard scales help space agencies and planetary defense organizations to prioritize monitoring and mitigation efforts based on the probability and severity of the potential impact.

The identification and tracking of near-Earth objects (NEOs) are essential components of asteroid impact modeling. By calculating the orbits of asteroids and predicting their trajectories, scientists can assess the likelihood of a collision with Earth. This process involves sophisticated astronomical observations and the use of computational tools to predict the asteroid's future position.

Many space agencies, such as NASA's Planetary Defense Coordination Office (PDCO), operate programs designed to track NEOs and calculate their trajectories. With the help of ground-based telescopes and space-based observatories, scientists monitor millions of objects in space, updating trajectory predictions as new data is gathered.

Asteroid tracking has become increasingly important in identifying high-risk objects and determining whether their orbits might intersect with Earth's. For example, the asteroid Apophis, which was once thought to have a small chance of hitting Earth in 2029, has been thoroughly studied, and its orbit has been refined over time to confirm that the risk is negligible.

In addition to statistical models and orbit tracking, scientists also use simulations to model the potential effects of asteroid impacts. These simulations are computer-based models that simulate the entire process of an asteroid collision, including the impact itself, the resulting shockwaves, and the long-term environmental effects.

1. Numerical Impact Simulations: These simulations calculate the energy released by an asteroid impact, the distribution of debris, and the resulting effects on Earth's climate and ecosystems. They are useful for understanding how large impacts would alter the planet's surface and atmosphere and can help identify the best methods for mitigating the damage.

2. Environmental Modeling: In addition to immediate effects, long-term environmental impacts are also modeled, including climate change, ecosystem disruption, and the potential for mass extinctions. These simulations incorporate data on the atmosphere, oceans, and terrestrial ecosystems to predict the extent of damage and the recovery timeline.

3. Planetary Defense Modeling: Finally, simulations are also used to test various mitigation strategies, including asteroid deflection techniques and evacuation plans. By modeling how different mitigation methods might affect an asteroid's trajectory and the impact outcome, scientists can determine the most effective strategies for planetary defense.

Modeling impact scenarios and calculating the probabilities of asteroid collisions are critical for understanding the potential risks posed by near-Earth objects. While asteroid impacts are relatively rare, the potential for global catastrophe means that ongoing research into asteroid detection, tracking, and mitigation remains essential. The use of statistical models, hazard scales, and impact simulations enables scientists to quantify risks and develop strategies to protect Earth from potential asteroid collisions. By continuing to improve these models, humanity will be better prepared to detect, track, and, if necessary, prevent a catastrophic impact in the future.

4.4. Near-Miss Events: What We've Learned

Near-omit occasions related to near-Earth items (NEOs) have turn out to be pivotal moments in humanity's growing consciousness of the cosmic dangers that orbit our planet. These close encounters—where asteroids or comets pass close sufficient to Earth to trigger subject but in the long run avoid collision—serve as essential case research in knowledge the dynamics of celestial our bodies, refining detection structures, and getting ready global response techniques. The instructions gleaned from those events have formed modern planetary defense and deepened our insight into both the vulnerabilities and resilience of our technological civilization.

Near-leave out activities are characterized through their potential to remind humanity of the skinny margin among normalcy and disaster. While many NEOs pass harmlessly, a few come inside some lunar distances or maybe closer, posing a tangible hazard that has sparked international scientific collaboration, media attention, and public problem. The increasing catalog of close to misses has found out patterns in asteroid populations, effect probabilities, and the effectiveness of early caution structures.

One of the earliest and maximum instructive close to misses turned into the close method of asteroid 2004 MN4, later named Apophis. Discovered in 2004, preliminary orbital calculations counseled a sizable risk of impact in 2029, sparking tremendous clinical and public alarm. Although further observations subtle its trajectory and ruled out an impending collision, Apophis served as a warning call, highlighting the significance of continuous monitoring, precise orbital monitoring, and fast communique. The event catalyzed investment and hobby in area-based totally commentary systems and global cooperation.

Near-misses offer valuable possibilities to test and enhance our detection technology. Ground-based totally telescopes, radar observations, and increasingly, spaceborne sensors have been hired to track those fast-shifting items with growing accuracy. Each near come upon allows astronomers to refine orbital fashions, take a look at the physical residences of NEOs, and calibrate instruments. For instance, radar imaging of close to-Earth asteroids at some point of close passes has found out details about their shape, rotation, and floor composition, which can be vital for developing effective mitigation techniques.

The observe of near misses has also underscored the significance of well timed and transparent conversation between scientists, governments, and the general public. Past occasions uncovered challenges in hazard assessment and message delivery, where uncertainty sometimes led to public anxiety or incorrect information. Lessons found out from these stories have knowledgeable protocols for conveying risk in ways that are scientifically accurate yet available, warding off panic whilst selling preparedness.

Moreover, close to-miss activities have fostered the development of worldwide frameworks for planetary defense. The establishment of organizations consisting of the Planetary Defense Coordination Office (PDCO) by NASA and the Space Missions Planning Advisory Group (SMPAG) through the United Nations exemplify institutional responses to the chance of impacts. These entities coordinate commentary campaigns, percentage statistics globally, and plan ability deflection or evacuation missions, integrating the scientific, political, and logistical dimensions of effect danger management.

Beyond technical and institutional improvements, near misses have motivated cultural perceptions of area dangers. Media coverage and public engagement at some stage in near approaches frequently convey the realities of cosmic threats into regular awareness. Documentaries, news reviews, and popular technology books use near-omit narratives to teach and inspire interest in astronomy and planetary defense. These cultural reflections enhance the urgency of preparedness and the marvel of dwelling on a dynamic planet within a extensive cosmic surroundings.

Scientific studies stimulated by way of close to misses also extends to expertise the broader population of NEOs, such as their origins and evolution. Close methods offer records that feed into fashions of ways those items are perturbed by using gravitational interactions, the Yarkovsky impact, and collisions, refining predictions of destiny trajectories. This know-how allows perceive capability future threats and informs the layout of missions geared toward asteroid characterization or deflection.

Another vital lesson from close to-leave out activities is the popularity of obstacles and uncertainties in modern-day detection skills. Some small asteroids have surpassed towards Earth than many satellites orbit, on occasion evading detection until hours before closest technique. These "marvel" close to misses emphasize the need for more suitable surveillance, together with the development of space-based totally infrared telescopes able to spotting dark, speedy-shifting objects against the backdrop of area.

The response to near misses has additionally highlighted the significance of contingency making plans at nearby, country wide, and global ranges. While most close to misses do no longer require immediately action beyond monitoring, the opportunity of an approaching hazard needs readiness to execute evacuation, effect mitigation, or disaster alleviation efforts. Exercises and simulations based totally on near-pass over eventualities help government and emergency responders prepare for real-international effect events.

Furthermore, close to misses have brought on philosophical and ethical reflections about humanity's region within the universe and responsibility to shield life on Earth. They raise questions on international fairness, because the consequences of impacts disproportionately have an effect on positive regions and populations. The task of coordinating a unified planetary defense attempt embodies the necessity of worldwide cooperation inside the face of shared existential threats.

Near-omit events have confirmed to be valuable training in celestial threat cognizance. They serve now not handiest as sensible exams of detection and reaction systems however additionally as profound reminders of the delicate boundary among safety and disaster. Through these near calls, humanity has superior its clinical know-how, delicate its technological gear, and reinforced its collective solve to monitor, apprehend, and in the long run mitigate the risks posed through near-Earth items.

CHAPTER 5

Asteroid Threat Mitigation Methods

5.1. Impulse Techniques and Deflection Solutions

In the context of planetary defense and asteroid deflection, nonnuclear impulse methods check with propulsion technology which can adjust the trajectory of an asteroid without using nuclear explosives. These techniques goal to divert the course of a probably risky asteroid (PHA) and decrease the hazard of impact with Earth. Here are a few nonnuclear impulse methods used for asteroid deflection:

Ion Propulsion: Ion propulsion is a exceptionally efficient and lowthrust propulsion system that uses electrically charged ions as propellant. It generates thrust by using accelerating ions using an electric powered discipline and expelling them thru a nozzle, growing a gentle and continuous acceleration. While ion propulsion affords low thrust, it could perform for prolonged periods, resulting in extensive pace modifications through the years. Spacecraft geared up with ion propulsion can progressively adjust an asteroid's trajectory by way of exerting a continuous pressure on the floor.

Solar Sail: A solar sail is a huge, thin, and enormously reflective sail that harnesses the momentum of photons from the Sun to generate thrust. As light from the Sun displays off the sail, it imparts momentum to the spacecraft, pushing it within the contrary path. Solar sails are especially effective inside the vacuum of area, wherein they can reap continuous acceleration over lengthy distances. For asteroid deflection, a spacecraft with a solar sail could be positioned close to the asteroid to utilize the strain of sunlight and exchange its trajectory over the years.

Gravity Tractor: The gravity tractor approach involves setting a spacecraft near a PHA and utilizing the gravitational appeal between the two bodies to steadily alter the asteroid's orbit. The spacecraft's gravitational have an impact on creates a tiny but non-stop acceleration at the asteroid. Over time, this small gravitational force can appreciably adjust the asteroid's direction and thoroughly divert it away from Earth. The gravity tractor approach calls for longduration near proximity to the asteroid but is an attractive choice due to its nondestructive nature.

Kinetic Impactors: While kinetic impactors contain a bodily collision with the asteroid, they're considered nonnuclear because they do no longer involve nuclear explosions. In this method, a spacecraft is deliberately directed to effect the asteroid at high pace. The effect imparts momentum to the asteroid, converting its pace and trajectory. The key difference from nuclear alternatives is that no nuclear explosive gadgets are used; rather, the spacecraft itself serves because the impactor.

Laser Ablation: Laser ablation involves the use of highpowered lasers to warmth the surface of an asteroid. The laser energy reasons the surface cloth to vaporize and create a jet of gas, which generates a thrust in the contrary path. By focused on unique areas on the asteroid's floor, scientists can observe thrust to change its path steadily.

Each of these nonnuclear impulse strategies has its benefits and obstacles, and the choice of method relies upon at the specific traits of the PHA and the available assets and time for the deflection mission. Scientists and space agencies maintain to analyze and increase these technologies to beautify our planetary protection abilities and be prepared for any ability destiny asteroid threats.

When planning a undertaking to deflect a Potentially Hazardous Asteroid (PHA) and adjust its trajectory, there are crucial considerations regarding orbit deflection and the interaction of the spacecraft with the asteroid's floor. These elements play a big function inside the achievement and effectiveness of asteroid deflection missions. Here are the important thing elements of orbit deflection and asteroid surface interplay:

Orbit Deflection Techniques: Various nonnuclear and nuclear techniques may be used to adjust the orbit of a PHA:

a. NonNuclear Techniques: Ion propulsion, sun sails, gravity tractor, kinetic impactors, and laser ablation are nonnuclear techniques for progressively converting the asteroid's trajectory through mild and continuous forces or particular affects.

b. Nuclear Techniques: Nuclear options involve the usage of nuclear explosives to provide a effective impulse to the asteroid. These strategies are being considered as lastresort options and are nonetheless within the theoretical level.

Impact Velocity and Angle: For kinetic impactors, the spacecraft must be directed to effect the asteroid with a particular pace and perspective to achieve the preferred orbit deflection. Accurate targeting is important to ensure the effect presents the specified alternate in the asteroid's speed.

Asteroid Composition and Surface Properties: The composition and floor properties of the asteroid are vital factors that have an impact on the deflection method. The presence of volatiles, porosity, and regolith thickness can affect the final results of an impact or surface interaction. Understanding these homes helps in designing powerful deflection strategies.

SpacecraftAsteroid Interaction: When a spacecraft processes an asteroid for deflection, near proximity operations are required. Precise navigation and manipulate are essential to avoid collisions or accidental adherence to the asteroid's floor because of its low gravity.

Deflection Monitoring: After making use of the deflection technique, continuous tracking of the asteroid's trajectory is crucial. This allows scientists to assess the effectiveness of the deflection and make vital adjustments if required.

Gravitational Keyholes: Gravitational keyholes are small regions in area near a celestial body where its gravity can drastically adjust the trajectory of an item passing through it. During asteroid deflection missions, averting gravitational keyholes is important to prevent unintended path changes.

Mission Timeframe: The undertaking time frame is important in deflection making plans. Early detection of a ability impact permits for an extended preparation duration, increasing the chance of a a hit deflection.

International Collaboration: Given the worldwide nature of the impact threat, worldwide collaboration is essential in deflection missions. Cooperation between area businesses and companies ensures shared resources, understanding, and information to address asteroid threats successfully.

Successfully deflecting a PHA requires careful consideration of those factors and the application of appropriate deflection strategies. As research and generation maintain to enhance, our capability to cope with ability asteroid effect dangers and shield our planet will enhance, in the end enhancing our planetary protection talents.

5.2. Use of Nuclear Explosions and Risks

In the context of planetary defense and asteroid impact mitigation, nuclear blast scenarios contain using nuclear explosives to divert or fragment a Potentially Hazardous

Asteroid (PHA) that poses a huge chance to Earth. While nuclear alternatives are considered as lastresort measures, they're being studied as capacity deflection techniques for sure types of PHAs. Here are the important thing components of nuclear blast scenarios and tracking:

Nuclear Blast Deflection Techniques: Nuclear blast deflection methods may be broadly categorised into two eventualities:

a. Standoff Burst: In this scenario, a nuclear explosive device is detonated at a safe distance from the asteroid's surface. The resulting blast produces a burst of energy that exerts a force on the asteroid, altering its trajectory. The aim is to exchange the asteroid's route with out immediately impacting it.

b. Surface Burst: In the surface burst state of affairs, the nuclear explosive device is intentionally positioned on or underneath the asteroid's floor. The detonation generates a shockwave which could fracture and disperse the asteroid, diverting its trajectory thru the resulting momentum change.

Factors Considered in Nuclear Scenarios: Before considering nuclear deflection, numerous factors are evaluated, together with:

a. Asteroid Properties: The length, composition, rotation charge, and inner shape of the asteroid are assessed to decide if a nuclear alternative is possible.

b. Location and Timing: The asteroid's function in its orbit and its proximity to Earth at the time of capability effect are important in making plans a nuclear deflection mission.

c. Environmental Risks: Potential dangers related to nuclear detonations, inclusive of the introduction of smaller risky fragments or capacity radioactive contamination, are taken into consideration.

d. International Cooperation: Any choice to install nuclear deflection might require worldwide cooperation and adherence to applicable treaties.

Simulation and Monitoring: Prior to implementing a nuclear deflection mission, substantial simulations are performed to recognize the effects of the blast at the asteroid and determine its capacity effectiveness. Additionally, cautious monitoring of the asteroid's trajectory and properties is essential to make sure the fulfillment of the deflection and mitigate any unintentional results.

International Space Missions: Given the complexities and worldwide implications of nuclear deflection, global collaboration and coordination are crucial. Multiple area agencies and organizations might probably be worried in making plans, tracking, and executing this kind of project.

Ethics and Legal Considerations: Nuclear deflection situations boost ethical questions and criminal concerns, given the capability environmental impact and global legal frameworks.

It's essential to reiterate that nuclear deflection is taken into consideration a lastresort alternative because of its complexities, capability risks, and moral issues. Nonnuclear deflection techniques, including ion propulsion, gravity tractor, and kinetic impactors, are favored and are actively studied and advanced for asteroid deflection missions. Nuclear scenarios are studied as a part of comprehensive planetary protection efforts, making sure we're prepared for diverse capability asteroid effect eventualities while prioritizing nonnuclear options on every occasion possible.

The prospect of using nuclear blast scenarios or every other asteroid deflection methods increases crucial ethical and legal considerations. Any plan to regulate the trajectory of a Potentially Hazardous Asteroid (PHA) must be conducted with cautious attention of the capacity results and adherence to worldwide legal frameworks. Here are the key ethical and criminal concerns associated with asteroid deflection:

Ethical Considerations:

Safety of Earth and Humanity: The number one ethical consideration in asteroid deflection is the safety of Earth and its inhabitants. The purpose is to save you or decrease the capacity catastrophic effect of a PHA on human populations and the environment.

Risk Mitigation vs. Risk Creation: Any deflection approach, which includes nuclear blast eventualities, should be evaluated in phrases of the overall risk it poses compared to the hazard of the asteroid effect itself. Decisionmakers have to weigh the potential environmental and human fitness dangers of a nuclear detonation in opposition to the risks of the asteroid's impact.

Minimizing Collateral Damage: In considering nuclear deflection methods, efforts ought to be made to reduce ability collateral harm, inclusive of growing smaller dangerous fragments or inflicting unintentional disruptions in area.

International Cooperation: Ethical issues amplify to the need for worldwide cooperation and transparency. Planetary defense efforts are global in nature, and collaboration amongst countries is important to successfully deal with asteroid threats.

Legal Considerations:

Space Law and Treaties: Any asteroid deflection challenge have to observe space regulation and applicable global treaties. The Outer Space Treaty, as an example, prohibits using nuclear weapons in space.

Liability and Responsibility: The accountable birthday celebration or events for the deflection venture ought to be certainly defined, and legal responsibility issues need to be addressed in case of any damaging consequences.

Informed Consent: If the deflection project includes international cooperation, informed consent from all participating countries should be acquired.

Sovereignty and Space Activities: Ethical and felony issues need to additionally do not forget issues of national sovereignty and jurisdiction in area activities.

Ethical DecisionMaking: Transparent, inclusive, and moral decisionmaking approaches are vital to make sure that every one stakeholders are worried within the discussions and that the satisfactory hobbies of humanity and the surroundings are prioritized.

Public Awareness and Engagement: As asteroid deflection missions have worldwide implications, public consciousness and engagement are crucial. Governments and space agencies need to speak overtly with the public approximately their intentions and efforts associated with planetary defense.

The ethical and criminal concerns in asteroid deflection are multifaceted and complex. As generation and understanding of planetary protection enhance, it is vital to balance the need for shielding Earth with ethical concepts, worldwide cooperation, and compliance with space regulation. Transparent and inclusive decisionmaking, guided by way of moral concerns and global cooperation, could be instrumental in efficiently addressing capability asteroid threats at the same time as upholding the values of protection, responsibility, and sustainability.

5.3 Mass Evacuation Strategies and Earth Protection Plans

In the face of an asteroid threat, mitigation efforts extend beyond simply preventing the impact itself. While asteroid deflection and destruction technologies are critical, the potential consequences of a large asteroid collision may render Earth uninhabitable for a significant period. In these extreme cases, mass evacuation strategies and comprehensive Earth protection plans will be necessary to safeguard humanity and other forms of life.

While many asteroid impacts are expected to be relatively small and their effects localized, a significant asteroid, especially one over 1 kilometer in diameter, could produce devastating consequences. A large impact could cause massive tsunamis, widespread fires, earthquakes, and a global winter triggered by dust and aerosols injected into the atmosphere, significantly disrupting the Earth's climate. Such catastrophic events could lead to mass extinctions and the collapse of ecosystems. Under these circumstances, evacuation becomes an essential part of humanity's survival strategy.

The global nature of the asteroid threat means that mass evacuation would need to involve extensive international cooperation, coordination of resources, and the development of infrastructure that can sustain large numbers of people over extended periods. The feasibility of such an effort depends on various factors, including the size of the asteroid, the impact location, and the available technology. As the risk of asteroid impact remains relatively low, but with potentially catastrophic consequences, a multi-layered protection strategy—ranging from impact mitigation to evacuation—would be necessary to ensure global safety.

Before diving into specific evacuation strategies, it is crucial to understand the types of asteroid threats that would necessitate such an evacuation. Impact scenarios can be categorized based on several factors, including asteroid size, impact location, and the resulting environmental effects.

The geographic location of the asteroid impact would significantly influence the evacuation strategy. If the asteroid were to hit an ocean or sea, the immediate concern would be the generation of tsunamis, which could cause massive flooding along coastlines. In this scenario, coastal regions would need to be evacuated swiftly, possibly with few hours of notice.

For an asteroid impact on land, the extent of damage would depend on the size and velocity of the asteroid. A large impact could lead to global consequences such as the release of dust and gases that block sunlight, triggering a "nuclear winter" scenario. In this case, evacuation efforts might not be limited to regional or national efforts but could require global coordination to move people to areas with more stable environmental conditions.

The size of the asteroid is a critical factor in determining the scale of the evacuation. Small asteroids might not cause significant global disruptions, while a large asteroid (several kilometers in diameter) could cause mass extinction-level effects. For smaller impacts, local evacuation would be sufficient to protect affected populations from immediate threats such as shockwaves, fires, and tsunamis. However, for larger impacts, long-term evacuation would be required to escape the climatic and environmental changes following the collision.

Asteroid threats may be detected years or decades in advance, allowing for careful planning and execution of evacuation strategies. Early detection systems such as space-based telescopes and ground-based observatories are crucial for providing early warning of potential impact events. With sufficient lead time, evacuation could be carried out in stages, focusing on the most vulnerable populations first. However, if an asteroid impact were detected with little warning, a rapid response would be required, and evacuation would need to be executed as quickly as possible.

Before any mass evacuation can occur, safe zones must be identified. These are areas that would be least affected by the impact and its subsequent environmental effects. For example, regions far from coastlines or those in high altitude areas may offer some protection from the immediate effects of an asteroid collision, such as tsunamis or heat blasts. These safe zones must be capable of supporting large populations with the necessary resources for survival.

Evacuation zones must also consider long-term viability. A safe zone is not just one that is geographically distant from the impact site but one that can sustain life in the aftermath of an asteroid impact. Areas that have access to fresh water, agricultural resources, and shelter from environmental extremes will be crucial.

Mass evacuation would require extensive transportation networks capable of moving large numbers of people quickly. In a worst-case scenario, transportation systems will need to be adapted to handle millions of people with limited resources. High-speed trains, planes, and ships would be essential to move people to safety. The logistics of such a large-scale movement of populations would require unprecedented coordination across countries and continents.

Evacuation strategies must also consider the capacity of infrastructure to support evacuees once they arrive in safe zones. Temporary shelters, food distribution, and medical care would all need to be established quickly to avoid chaos and prevent collapse of the evacuee systems.

Given the global nature of the threat, mass evacuation strategies cannot be carried out by individual nations in isolation. An asteroid impact could affect every country on Earth, so international cooperation will be necessary to coordinate evacuation efforts. The United Nations or a similar global organization could serve as a central coordinating body, working alongside national governments, military organizations, and humanitarian agencies.

In addition to coordination, communication is critical. Real-time data on asteroid trajectories, expected impact zones, and safe zones will need to be continuously updated and shared between all nations. This data will help ensure that evacuation efforts are carried out efficiently and that people are directed to the most suitable areas for protection.

While the goal of evacuation is to protect as many people as possible, resources will inevitably be limited. In such a scenario, priority must be given to the most vulnerable populations, including the elderly, children, pregnant women, and individuals with special needs. Efforts will also need to focus on maintaining social order during the evacuation process, preventing panic, and ensuring equitable distribution of resources.

Once evacuees are moved to safe zones, long-term sustainability must be considered. With the potential for a protracted period of global environmental disruption, safe zones must have the capacity to provide food, water, and shelter for extended periods. Additionally, healthcare, sanitation, and energy supplies must be ensured to avoid disease outbreaks and infrastructure collapse.

Evacuation is not just about escaping the immediate effects of an asteroid impact; it is about ensuring that humanity can continue to thrive in the aftermath of the disaster. This requires the development of self-sustaining ecosystems and closed-loop systems capable of maintaining human life for years, if not decades, while the Earth recovers.

In addition to evacuation, Earth protection plans are necessary to ensure the long-term survival of life on Earth following an asteroid impact. These plans may include efforts to:

1. Restore Ecological Balance: Following a large impact, ecosystems will likely be severely disrupted. Efforts to restore biodiversity and stabilize the environment will be critical in the long-term recovery process. This could involve reintroducing species to affected areas, restoring habitats, and addressing soil and water contamination caused by the impact.

2. Planetary Defense Initiatives: While evacuation is a crucial component of survival, the best solution is to prevent an asteroid impact from occurring in the first place. Planetary defense technologies, including asteroid deflection methods, nuclear disruption strategies, and gravitational tractor techniques, must be developed and deployed to prevent a future catastrophe.

3. Space Colonization: In the most extreme cases, where Earth becomes too inhospitable to support life, space colonization could serve as a last resort. This would involve the establishment of human colonies on other planets or moons, utilizing advanced space technologies to support human life

away from Earth. While this is a long-term solution, it could serve as a safeguard against extinction in the face of an asteroid threat.

Mass evacuation strategies and Earth protection plans are vital components of a comprehensive asteroid threat mitigation strategy. While preventing an impact through asteroid deflection remains the most desirable solution, preparations for large-scale evacuation and long-term survival in the aftermath of an asteroid collision are necessary to ensure humanity's survival. These strategies require international cooperation, advanced infrastructure, and significant investments in technology and resources. As the threat of asteroid impacts, though rare, remains real, developing and refining these strategies is crucial for the future of human civilization.

5.4. Role of Private Aerospace Companies in Defense

In the evolving panorama of planetary defense, personal aerospace organizations have emerged as essential gamers along governmental area businesses. The developing threats posed by close to-Earth objects (NEOs) and other cosmic risks have necessitated a collaborative approach that leverages the innovation, agility, and assets of private enterprise. The involvement of private aerospace corporations in defense initiatives marks a good sized shift from historically state-dominated area exploration and hazard mitigation efforts, underscoring the expanding function of the industrial sector in securing Earth's future.

Historically, planetary protection changed into normally the domain of countrywide space corporations which includes NASA, ESA, and Roscosmos. These businesses undertook the huge duties of tracking doubtlessly risky asteroids, growing deflection techniques, and carrying out medical research. However, the rapid improvements in aerospace era, coupled with elevated investment in commercial spaceflight, have catalyzed a brand new technology where private agencies contribute not most effective technology and knowledge but also strategic vision and monetary assets.

One of the foremost contributions of private aerospace corporations lies within the development and deployment of modern technologies for detection and monitoring of NEOs. Companies specializing in satellite production, launch offerings, and space remark devices offer the hardware vital for comprehensive monitoring networks. For instance, private firms have pioneered the introduction of small satellite tv for pc constellations able to close to-non-stop surveillance of the sky, notably augmenting the information amassed through traditional observatories.

Furthermore, personal aerospace companies are at the leading edge of task layout and execution for lively defense measures. Concepts such as kinetic impactors, gravity tractors, and even extra speculative technologies like laser ablation require sophisticated spacecraft capable of precise navigation and control. The knowledge of business area firms in spacecraft development, propulsion structures, and project management is valuable in turning these theoretical strategies into practical solutions. Notably, some corporations have introduced bold plans to participate at once in planetary defense missions, collaborating with government organizations to layout and release asteroid deflection probes.

The agility of private aerospace organizations allows quicker innovation cycles and reduced improvement fees in comparison to standard authorities projects. Competitive pressures drive those companies to optimize engineering procedures, undertake new substances, and implement advanced software program systems. This efficiency speeds up the deployment of recent abilities important for early detection and speedy reaction to emerging threats. Moreover, non-public region participation facilitates diverse approaches, as multiple companies pursue various technologies and task architectures, increasing the general robustness of planetary protection efforts.

Financing is every other domain in which non-public aerospace organizations impact planetary protection. Through project capital, public services, and partnerships, these corporations mobilize extensive investment in the direction of space technology that could have dual applications in defense and commercial sectors. The infusion of personal capital complements public funding, enabling extra bold projects and sustained development. Additionally, non-public organizations' involvement attracts international interest and public hobby, that could translate into broader guide for planetary protection tasks.

The position of private aerospace corporations extends beyond era improvement to encompass records analysis, software structures, and statistics sharing. Many groups develop state-of-the-art algorithms for orbital prediction, effect risk assessment, and venture simulation. By providing cloud-based totally platforms and open information tasks, they decorate collaboration amongst scientists, policymakers, and emergency managers global. This integration of records-pushed approaches improves the accuracy and timeliness of danger evaluation, which is crucial for powerful decision-making in disaster situations.

International collaboration is every other place prompted with the aid of personal aerospace agencies. While planetary defense is inherently international, related to more than one international locations and corporations, non-public firms operate across borders, forging partnerships that go beyond traditional geopolitical limitations. These collaborations foster expertise change, harmonize technical standards, and create incorporated response frameworks. The participation of personal entities in global running corporations and advisory committees guarantees that business views and skills inform global techniques.

Despite their growing impact, personal aerospace corporations face demanding situations in absolutely integrating into planetary defense frameworks. Regulatory hurdles, liability concerns, and the want for coordination with national protection businesses require cautious navigation. Transparency and accountability are important, as protection-associated sports should align with international treaties and moral requirements. Governments and private firms continue to barter the terms of collaboration, striving to stability industrial pursuits with public safety imperatives.

Public notion and communique also play a function in shaping the involvement of private aerospace groups. High-profile missions and improvements capture imaginations and encourage guide however additionally improve expectations and scrutiny. Effective outreach and education are vital to bring the complexities and barriers of planetary defense technology, fostering informed public discourse. Private groups more and more have interaction in technological know-how communique and partnership with instructional institutions to construct trust and recognition.

Looking to the destiny, the function of personal aerospace corporations in planetary protection is poised to extend. Emerging technologies including synthetic intelligence, self sufficient spacecraft, and superior propulsion systems promise to revolutionize detection and mitigation abilities. Private corporations are properly-located to pioneer those improvements, pushing the boundaries of what's possible. Moreover, the prospect of business asteroid mining and space useful resource utilization offers synergies with protection targets, as infrastructure advanced for financial functions may be leveraged for planetary safety.

Private aerospace companies have come to be crucial members to planetary defense, complementing governmental efforts with innovation, efficiency, and financial sources. Their involvement enriches the technological landscape, complements global collaboration, and strengthens humanity's ability to count on and respond to cosmic threats. As the challenges of planetary protection grow in scale and complexity, the partnership among public companies and private industry may be primary to securing Earth's safety within the cosmic surroundings.

CHAPTER 6

Advanced Technologies and Space Missions

6.1. Space Observation and Exploration Vehicles

Space telescopes and observation satellites play a important position in the detection, monitoring, and look at of celestial gadgets, such as Potentially Hazardous Asteroids (PHAs) and different nearEarth items (NEOs). These superior gadgets are designed to study the universe from area, presenting valuable information for medical research, planetary protection, and area exploration. Here are the key aspects of space telescopes and remark satellites:

Detection of PHAs: Space telescopes are ready with sensitive detectors capable of scanning large regions of the sky looking for PHAs and NEOs. They can discover faint and distant objects that might be missed by using groundbased observatories because of atmospheric interference.

Early Warning Systems: Spacebased remark satellites make a contribution to early caution structures for ability asteroid influences. By detecting and monitoring PHAs properly earlier, scientists can as it should be expect their trajectories and assess their capability effect risks on Earth.

Precise Orbit Determination: Observation satellites prepared with highresolution imaging and spectroscopic capabilities permit scientists to precisely decide the orbits of PHAs. Accurate orbit dedication is vital for predicting destiny close strategies and potential impact events.

Continuous Monitoring: Space telescopes in orbit provide continuous monitoring of PHAs and different celestial gadgets. They can take a look at objectives without the constraints of daytime or atmospheric conditions, ensuring non-stop monitoring of asteroids around the clock.

NearInfrared and Thermal Observations: Many space telescopes are ready with nearinfrared and thermal imaging abilties. These observations permit scientists to have a look at the surface homes, composition, and thermal traits of asteroids, imparting insights into their shape and conduct.

Surveys and Cataloging: Space telescopes are used to behavior complete surveys of the sky, contributing to the cataloging of asteroids and NEOs. These surveys help become aware of potential effect threats and provide vital statistics for planetary protection efforts.

SpaceBased Collaboration: International collaboration is commonplace in area exploration and remark. Various area agencies and companies around the world work together to release and function space telescopes and commentary satellites, pooling sources and expertise to increase our information of asteroids and different celestial items.

Future Missions: As technology advances, new area telescopes and observation satellites are deliberate to beautify our skills for detecting, monitoring, and studying PHAs and NEOs. Upcoming missions aim to enhance our capability to pick out and signify ability impact threats and aid planetary defense initiatives.

Data Sharing and Open Science: The data collected by means of space telescopes and observation satellites are usually shared brazenly with the worldwide scientific community. This exercise fosters collaborative studies and ensures that a diverse variety of specialists can examine and interpret the observations, furthering our know-how of asteroids and their ability impact dangers.

Space telescopes and commentary satellites have revolutionized our information of the cosmos, which include the detection and have a look at of PHAs and NEOs. Their non-stop monitoring and unique observations contribute drastically to planetary protection efforts, providing important statistics for assessing capability impact dangers and informing deflection strategies if essential.

Space mining and useful resource utilization consult with the concept of extracting and utilizing precious materials and sources from celestial bodies, which include asteroids, the Moon, and different planets. This rising area of space exploration holds the capacity for helping destiny area missions, advancing space generation, or even reaping rewards existence on Earth. Here are the important thing components of space mining and useful resource usage:

Asteroid Mining: Asteroids are rich in various valuable assets, together with metals like iron, nickel, and platinumgroup metals, in addition to water and risky elements. Space mining pursuits to extract those resources from asteroids to be used in area missions or transported back to Earth for various purposes.

Lunar Resource Utilization: The Moon is another goal for aid usage. Lunar regolith, the free layer of soil and dust masking the Moon's floor, carries valuable elements like oxygen, silicon, and aluminum. These resources may be used for building systems and supporting future human settlements on the Moon or in space.

InSitu Resource Utilization (ISRU): The concept of insitu resource usage includes using domestically available materials on celestial bodies to support human activities. ISRU reduces the want to transport huge amounts of assets from Earth, making space missions more sustainable and costeffective.

Propellant Production: One of the main desires of useful resource usage is to supply propellants, which includes liquid oxygen and hydrogen, from available sources. Propellant manufacturing in space would lessen the cost and complexity of space missions by using putting off the want to transport huge amounts of propellant from Earth.

Space Infrastructure: Resource usage can help the development of space infrastructure, consisting of space habitats, bases, and fueling stations. These centers should serve as stepping stones for further exploration and colonization of area.

Space Manufacturing: Resources extracted from celestial our bodies may be used for space production, growing merchandise and components that would otherwise be hard to transport from Earth. This capability ought to revolutionize area industries and enable new medical and commercial activities.

Sustainability and Space Exploration: Resource utilization is seen as a crucial thing of sustainable space exploration. By harnessing local assets, space missions can grow to be selfsustaining and less reliant on Earth for vital resources.

Technological Challenges: Space mining and useful resource usage pose great technological challenges. These include designing efficient extraction strategies, coping with and processing materials in lowgravity environments, and growing secure and automated mining systems.

Legal and Ethical Considerations: The criminal and ethical implications of space mining and useful resource utilization are complicated and want to be addressed. Issues related to assets rights, environmental effect, and equitable useful resource sharing amongst international locations are a number of the considerations.

Space mining and useful resource usage preserve brilliant promise for advancing area exploration and assisting future space missions. While challenges exist, ongoing research and improvement in this subject have the capability to convert area activities and pave the manner for sustainable human presence past Earth. However, it's miles critical to cope with the related felony, ethical, and technological demanding situations to ensure accountable and useful space useful resource usage.

6.2. CuttingEdge Technologies and Research on Asteroids

Highprecision monitoring and modeling structures are crucial additives of planetary defense efforts, in particular in terms of monitoring Potentially Hazardous Asteroids (PHAs) and predicting their destiny trajectories. These advanced structures allow scientists to accurately tune the positions and actions of asteroids, permitting precise orbit determination and impact chance assessments. Here are the important thing aspects of highprecision tracking and modeling systems:

Radar and Optical Telescopes: Both radar and optical telescopes are used for monitoring PHAs. Radar telescopes can have a look at asteroids day and night time, unaffected with the aid of daylight or atmospheric situations, and provide specific statistics approximately an asteroid's length, shape, rotation, and distance. Optical telescopes use seen mild to observe asteroids and are also valuable in tracking their positions over the years.

Astrometry and Ephemeris: Astrometry entails measuring the best positions of celestial objects. In the context of PHAs, astrometry is used to as it should be determine the positions of asteroids at exceptional factors in time. The resulting information is used to create ephemeris tables, which give expected positions of asteroids within the destiny.

Radar DelayDoppler Tracking: Radar delayDoppler monitoring is a specialized method that uses radar alerts bounced off asteroids to measure their distances and velocities correctly. This technique gives special data about an asteroid's movement and orbit.

Orbit Determination and Uncertainty: Highprecision monitoring records is used along with sophisticated numerical algorithms to determine the orbits of PHAs. These algorithms recall gravitational interactions with other celestial bodies and decrease uncertainties in the expected orbits.

Observational Campaigns: To improve the accuracy of orbital predictions and impact hazard tests, dedicated observational campaigns are performed for newly discovered PHAs or those which might be approaching Earth intently. These campaigns contain a couple of observatories round the sector to acquire redundant information.

Computational Models and Simulations: Computational fashions and simulations are used to expect the destiny trajectories of PHAs based totally on monitoring records and known physical legal guidelines. These simulations allow scientists to evaluate the probability of capacity effect activities and plan for deflection missions if necessary.

NearEarth Object Coordination and Databases: International companies, which include the International Astronomical Union's Minor Planet Center (MPC) and the European Space Agency's NearEarth Object Coordination Centre (NEOCC), coordinate tracking efforts and preserve databases of PHAs and NEOs.

Early Warning Systems: Highprecision tracking systems play a important position in early caution structures, offering enough time to assess capacity effect risks and take suitable movements in the occasion of a risky asteroid.

Continuous Monitoring: Tracking systems permit continuous monitoring of PHAs, permitting scientists to have a look at changes in their orbits and refine effect risk exams as new information will become available.

Highprecision tracking and modeling systems are integral to planetary defense, presenting critical facts for information the orbits of PHAs and accurately predicting capacity effect occasions. These advanced technologies and observational efforts play a essential position in safeguarding Earth from the potential hazards posed through nearEarth objects.

Mission making plans and trajectories are essential additives of planetary defense efforts while handling Potentially Hazardous Asteroids (PHAs). These involve cautious concerns and calculations to layout deflection missions that effectively adjust the trajectory of a threatening asteroid and mitigate potential effect risks. Here are the important thing components of challenge planning and trajectories for PHA deflection:

Early Detection and Assessment: Early detection of PHAs is vital to permit enough time for undertaking making plans and coaching. Once a doubtlessly hazardous asteroid is diagnosed, scientists conduct distinct tests of its length, composition, orbit, and capacity effect dangers.

Orbital Analysis and Prediction: Precise orbital evaluation of the PHA is finished based on observational records from groundbased and spacebased telescopes. This evaluation allows expect the asteroid's future trajectory, including its potential close processes to Earth and viable effect situations.

Impact Probability Calculation: By combining the orbital analysis with regarded uncertainties in the asteroid's trajectory, scientists calculate the impact possibility over numerous destiny time frames. This facilitates determine the urgency and priority of the deflection mission.

Deflection Techniques: Based on the characteristics of the PHA and the time to be had for deflection, scientists examine diverse deflection strategies, which include nonnuclear alternatives like gravity tractor, kinetic impactor, ion propulsion, and nuclear options like standoff burst or surface burst.

Mission Timeframe and Targeting Window: The time to be had for the deflection challenge is a critical issue. Depending on the orbital positions of Earth and the asteroid, there is probably specific home windows of possibility for a a success deflection. Mission planners should optimize the timing to gain the preferred trajectory exchange.

Spacecraft Design and Payload: The design of the spacecraft and the payload used for deflection are essential. Spacecraft need to carry the vital propulsion systems, gadgets for navigation and targeting, and any deflection payloads required to regulate the asteroid's trajectory.

Trajectory Planning: Mission planners calculate the trajectory of the spacecraft, taking into consideration gravitational forces, propulsion maneuvers, and any potential flybys of other celestial our bodies. The trajectory wishes to be correctly planned to make sure the spacecraft reaches the goal asteroid and performs the deflection maneuver as supposed.

Gravity Assist and Flyby Opportunities: Mission planners search for opportunities to use gravity assists or flybys of different celestial our bodies to help finetune the spacecraft's trajectory and conserve gasoline in the course of the undertaking.

Monitoring and Adjustments: Throughout the project, non-stop tracking of the spacecraft and the asteroid is essential. Scientists make any required trajectory changes to make sure the spacecraft stays on course and performs the deflection maneuver accurately.

Contingency Planning: Mission planners put together for contingencies and sudden consequences. Plans are in region to evolve to new facts or unexpected challenges which could rise up for the duration of the project.

Mission making plans and trajectories are crucial factors in correctly deflecting Potentially Hazardous Asteroids. By combining precise orbital analysis, deflection technique choice, spacecraft design, and non-stop monitoring, scientists can devise powerful deflection missions to guard our planet from ability effect hazards. Early detection and evaluation of PHAs are key to supplying sufficient time for mission making plans and execution, maximizing the chances of a successful deflection and minimizing the ability dangers related to unsafe asteroids.

6.3 Upcoming Asteroid Missions and Their Potential

As humanity faces the persistent threat of asteroid collisions, space agencies and research organizations across the globe are investing in new technologies and missions to study, track, and ultimately prevent catastrophic asteroid impacts. Over the past few decades, space missions have greatly advanced our understanding of asteroids, including their composition, behavior, and potential risks. However, as we move into the future, several groundbreaking missions are set to launch that could dramatically enhance our ability to detect, characterize, and even deflect asteroids that might pose a danger to Earth.

NASA's Psyche Mission

One of the most highly anticipated asteroid missions is NASA's Psyche mission, set to launch in 2023. This mission aims to explore the asteroid 16 Psyche, which is unique in that it is composed largely of metallic iron and nickel, similar to Earth's core. Psyche is believed to be the remnant core of a planetesimal that underwent a collision early in the solar system's history, offering scientists a unique opportunity to study the building blocks of planetary cores.

While Psyche's primary objective is to gain insight into the formation of planets, it also has major implications for planetary defense. By understanding the composition and characteristics of different types of asteroids, scientists can better predict the behavior of potentially hazardous asteroids and develop effective deflection methods. Psyche's study could lead to a deeper understanding of asteroid impacts and their potential effects on Earth, helping to refine models for impact scenarios and mitigation strategies.

Mission Highlights:
- Target: Asteroid 16 Psyche
- Launch: 2023

- Objective: Study the metallic asteroid to understand the composition of planetary cores.
- Potential: Insights into the composition of asteroid materials can inform planetary defense strategies and help develop deflection techniques.

ESA's Hera Mission

The European Space Agency (ESA) is planning to launch the Hera mission as part of the AIDA (Asteroid Impact & Deflection Assessment) collaboration with NASA's DART (Double Asteroid Redirection Test) mission. Hera's primary goal is to study the binary asteroid system Didymos, which consists of a large primary asteroid and a smaller moonlet, known as Didymoon or Didymos B. This pair offers an excellent opportunity for testing asteroid deflection technologies.

In 2022, NASA's DART mission made history by successfully impacting Didymoon with the goal of altering its orbit, which will be measured by Hera. Hera will provide key data on how asteroids respond to impact and deflection, offering critical insight into the effectiveness of these planetary defense methods. This mission will mark the first time that humanity tests the ability to change the trajectory of an asteroid, which could potentially be used to divert dangerous asteroids from Earth's path in the future.

Mission Highlights:

- Target: Didymos (binary asteroid system)

- Launch: 2024 (Hera mission)
- Objective: Assess the effects of impact on an asteroid and refine planetary defense methods.
- Potential: Help determine the effectiveness of kinetic impactor techniques for deflecting dangerous asteroids.

OSIRIS-REx Sample Return Mission

NASA's OSIRIS-REx mission, which launched in 2016, was designed to collect samples from the near-Earth asteroid Bennu. This asteroid, which has been identified as a potentially hazardous object, is expected to have a 1-in-2,700 chance of impacting Earth over the next 200 years. The mission aims to retrieve samples from Bennu's surface and return them to Earth in 2023.

OSIRIS-REx has provided scientists with invaluable data on the asteroid's size, shape, rotation, and surface characteristics. These findings will inform impact risk assessments and improve our understanding of how objects like Bennu might behave in a collision scenario. Furthermore, the samples collected will allow scientists to study the asteroid's composition, including the presence of organic molecules and water, offering clues to the origins of life and the conditions in the early solar system.

The OSIRIS-REx mission's findings will help refine planetary defense strategies by improving our understanding of the structure and materials that make up hazardous asteroids. By studying Bennu's surface properties, researchers can develop more accurate models for predicting how asteroids might react to various deflection techniques.

Mission Highlights:

• Target: Asteroid Bennu

• Launch: 2016 (return of samples expected in 2023)

• Objective: Collect and return samples from a potentially hazardous asteroid to study its composition and origins.

• Potential: Provide critical data on asteroid composition and surface characteristics to aid in impact prediction and mitigation.

Japan's Hayabusa2 Mission

Japan's Hayabusa2 mission, which successfully returned samples from the asteroid Ryugu in December 2020, is another milestone in asteroid research. Ryugu, like Bennu, is considered a primitive asteroid, rich in materials that are thought to have existed in the early solar system. By studying these samples, scientists can learn more about the chemical processes that led to the formation of the solar system and gain insights into the possibility of water and organic molecules being delivered to Earth by asteroids.

The Hayabusa2 mission's success in collecting samples has already yielded valuable information that will inform future planetary defense efforts. The data on Ryugu's surface material, for instance, could help researchers better understand the structural integrity of asteroids and how they might respond to deflection techniques such as kinetic impacts or nuclear explosions.

Mission Highlights:

- Target: Asteroid Ryugu
- Launch: 2014 (samples returned in 2020)
- Objective: Collect and return samples to study the early solar system and organic compounds.
- Potential: Enhance understanding of asteroid composition and behavior to improve deflection strategies.

China's Tianwen-1 and Asteroid Exploration Plans

China has made significant strides in its space exploration efforts, and the country is planning its own asteroid missions in the coming years. While the primary focus of the Tianwen-1 mission, which successfully landed a rover on Mars in 2021, was Mars exploration, China has set its sights on asteroid exploration as part of its broader planetary defense strategy.

The Chinese space agency is planning missions to explore and study near-Earth objects, with a long-term goal of developing technologies for asteroid deflection and resource utilization. These missions are expected to use advanced spacecraft capable of landing on asteroids, collecting samples, and providing detailed data on their composition and trajectory.

Mission Highlights:

- Target: Near-Earth objects (NEOs)

- Launch: Planned for the 2020s and beyond
- Objective: Study the composition and potential hazards posed by NEOs, along with developing deflection technologies.
- Potential: Contribute to global asteroid tracking systems and provide data for planetary defense technologies.

The Role of AI and Machine Learning in Asteroid Missions

One of the most promising aspects of upcoming asteroid missions is the integration of artificial intelligence (AI) and machine learning into space exploration technologies. AI and machine learning algorithms can analyze vast amounts of data from telescopes, spacecraft, and asteroid observations in real time, improving our ability to detect and track potentially hazardous objects.

AI technologies are particularly useful in predicting asteroid trajectories and assessing the potential impact risk posed by near-Earth objects. These systems can also assist in optimizing mission planning and spacecraft navigation, enabling more precise and efficient asteroid exploration.

Mission Highlights:
- Role of AI: Real-time data analysis, trajectory prediction, and mission optimization.
- Impact: Enhance asteroid detection capabilities and improve planetary defense efforts by using advanced algorithms to track and model asteroid behavior.

Upcoming asteroid missions are poised to revolutionize our understanding of these ancient cosmic objects, offering vital insights into the risks they pose to Earth. Through innovative space missions such as NASA's Psyche, ESA's Hera, and Japan's Hayabusa2, we are gaining critical knowledge about asteroid composition, behavior, and potential impact scenarios. As technology advances, missions utilizing AI and machine learning will further enhance our ability to track and assess asteroid risks, ultimately contributing to the development of effective planetary defense strategies. These upcoming missions hold immense potential for improving our preparedness for asteroid threats and ensuring the long-term survival of humanity.

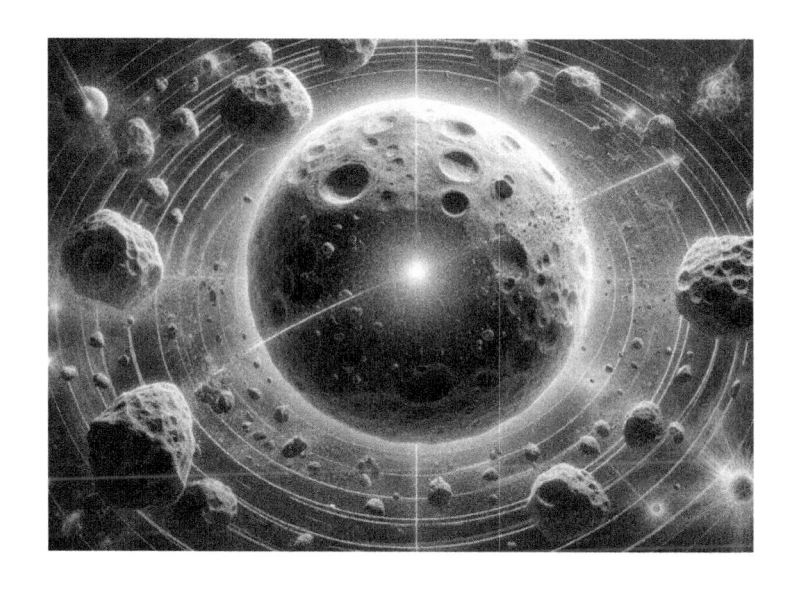

CHAPTER 7

Crisis Management and Public Awareness

7.1. Crisis Management all through Potential Threats

International Asteroid Warning Network (IAWN): IAWN is a global collaboration of observatories, space companies, and institutions committed to sharing statistics and coordinating efforts to come across, music, and are expecting the trajectories of Potentially Hazardous Asteroids (PHAs) and other NearEarth Objects (NEOs). It makes a speciality of early warning and danger assessment to inform and prepare decisionmakers and the public for ability effect activities.

Space Data Coordination Group (SDCG): SDCG is a operating organization set up under the United Nations Committee on the Peaceful Uses of Outer Space (COPUOS). It helps the alternate of records and records related to area objects, consisting of NEOs. Its task is to enhance the sharing of data on PHAs and sell international collaboration in planetary defense.

United Nations Committee on the Peaceful Uses of Outer Space (COPUOS): COPUOS is a UN frame liable for selling global cooperation within the non violent makes use of of outer space. It addresses diverse spacerelated troubles, which include planetary protection. Member states collaborate to establish tips and protocols for international cooperation in detecting and mitigating asteroid effect threats.

European Space Agency (ESA): ESA is an intergovernmental business enterprise committed to area exploration and studies. ESA conducts missions associated with planetary protection and space remark, contributing precious facts and know-how to global efforts.

National Aeronautics and Space Administration (NASA): NASA, the distance organisation of the United States, plays a good sized function in asteroid detection, tracking, and planetary protection. It leads diverse missions, which includes the NearEarth Object Observations Program, aimed toward coming across, characterizing, and tracking NEOs.

Japan Aerospace Exploration Agency (JAXA): JAXA, the distance agency of Japan, conducts area missions associated with planetary protection and contributes to worldwide efforts in asteroid studies and observation.

NearEarth Object Observations Program (NEOO) USA: The NEOO software is led by means of NASA and is chargeable for discovering, monitoring, and characterizing NEOs. It plays a vital role in presenting early caution of capability effect threats and assessing effect risks.

The Torino Scale and the Palermo Scale: The Torino Scale and the Palermo Scale are protocols used to speak the ability impact risks of NEOs, inclusive of PHAs. They provide standardized scales for assessing and conveying the seriousness of feasible effect activities to the general public and decisionmakers.

International Collaboration Agreements: Various space groups and groups have bilateral or multilateral agreements for cooperation in asteroid research and planetary defense. These agreements facilitate data sharing, joint missions, and the coordination of efforts to guard the Earth from potential asteroid affects.

International companies and protocols play a important function in planetary defense efforts related to PHAs and NEOs. Through collaboration and records sharing, those entities work together to stumble on, song, and verify capability effect risks and develop techniques to defend our planet from capacity asteroid threats. The collective effort of those groups ensures a coordinated and international response to the demanding situations posed with the aid of Potentially Hazardous Asteroids.

Rapid reaction plans and communique techniques are vital factors in the context of planetary protection, specifically whilst managing the potential effect of Potentially Hazardous Asteroids (PHAs). These plans and techniques are designed to ensure well timed and effective coordination amongst relevant groups and authorities in the occasion of a detected PHA with a vast effect danger. Here are the key aspects of rapid response plans and communique techniques:

Early Warning and Alert Systems: Rapid response plans consist of early warning and alert structures that permit the timely detection of PHAs and capability impact activities.

These systems use advanced telescopes, radar, and data analysis algorithms to pick out and verify doubtlessly hazardous gadgets.

Impact Probability Assessment: Once a PHA is detected, effect chance assessment is performed to evaluate the likelihood of an impact. Sophisticated numerical simulations and statistical analyses assist estimate the opportunity and potential impact locations.

Coordination amongst Space Agencies and Observatories: International and regional space businesses, observatories, and studies establishments collaborate to percentage statistics and coordinate efforts. Rapid response plans outline clean communique channels and obligations among these entities to make certain efficient reaction coordination.

Scenario Analysis and Mission Planning: Rapid reaction plans contain situation evaluation to assess specific ability consequences based on the available data. Mission planning for deflection missions or different mitigation strategies is also taken into consideration if the impact hazard is good sized.

Emergency Response Protocols: In case of a excessive impact possibility, rapid reaction plans encompass emergency response protocols to be followed by means of applicable government, governments, and catastrophe control organizations. These protocols outline movements to be taken to limit the effect's potential outcomes.

DecisionMaking Processes: Clear decisionmaking procedures are mounted in rapid response plans to manual authorities in figuring out the right path of motion. This includes considering the uncertainties related to the asteroid's trajectory and the effectiveness of various deflection alternatives.

Public Communication Strategies: Effective communication with the public is a vital factor of speedy response plans. Communication strategies are designed to offer correct and timely data to the public, addressing capacity risks, mitigation plans, and emergency approaches.

Media and International Cooperation: Rapid reaction plans consist of techniques for engaging with the media to ensure correct reporting and avoid incorrect information. They additionally facilitate global cooperation and statistics exchange to cope with the global implications of ability impact events.

Tabletop Exercises and Drills: To test the effectiveness of fast response plans, tabletop sporting events and drills are conducted concerning relevant stakeholders. These simulations help perceive ability weaknesses inside the response method and improve usual preparedness.

Continual Review and Updates: Rapid reaction plans are situation to persistent overview and updates primarily based on technological improvements, new studies, and lessons discovered from past occasions. Regular reviews make certain that the plans continue to be uptodate and powerful.

Rapid response plans and conversation techniques are critical in preparing for and addressing potential impact events of Potentially Hazardous Asteroids. These plans make sure coordinated and efficient responses, early warning, powerful decisionmaking, and accurate verbal exchange with the general public. By combining global cooperation, advanced era, and emergency preparedness, the worldwide community can higher protect towards the effect dangers posed by means of PHAs.

7.2. Public Awareness and Education

Public awareness and schooling are essential components of planetary defense efforts to cope with the capability effect of Potentially Hazardous Asteroids (PHAs) and other NearEarth Objects (NEOs). Effective verbal exchange tools are used to inform and have interaction the general public, enhance recognition approximately asteroid effect dangers, and sell preparedness. Here are the key conversation gear used to growth public focus:

Public Websites and Portals: Dedicated public web sites and portals are created by way of space organizations, observatories, and agencies worried in planetary protection. These structures provide smooth get right of entry to to statistics approximately PHAs, NEOs, and effect risks. They often consist of educational resources, interactive features, and realtime updates on asteroids' trajectories.

Social Media Channels: Social media platforms, along with Twitter, Facebook, and Instagram, play a sizable position in achieving a wide target market. Space agencies and organizations use these channels to percentage news, academic content, visualizations, and impactrelated updates. Engaging content material is shared to spark hobby and encourage public participation.

Public Lectures and Workshops: Public lectures, workshops, and seminars are prepared with the aid of area groups, astronomy golf equipment, and academic establishments to inform the general public about asteroid threats and planetary protection efforts. These occasions characteristic expert speakers and permit attendees to ask questions and advantage insights into the problem.

Educational Outreach Programs: Schools and educational institutions are involved in academic outreach applications that concentrate on planetary protection and asteroids. Space agencies collaborate with educators to expand ageappropriate curricula and handson activities to educate students about asteroids, impact dangers, and mitigation strategies.

Mobile Apps and Games: Interactive cell apps and video games associated with asteroid effect situations are evolved to engage the public, specially younger audiences. These apps permit customers to explore asteroids, simulate impact occasions, and learn about planetary defense in an exciting and academic way.

Infographics and Visualizations: Infographics and visualizations are powerful tools to present complicated records in a visually attractive and easytounderstand way. These visuals illustrate asteroid effect chances, capability results, and the significance of preparedness.

Public Service Announcements (PSAs): Television and radio PSAs are used to deliver concise messages about asteroid impact dangers and the importance of staying knowledgeable and organized. These brief bulletins increase focus amongst the general public and inspire them to are trying to find more statistics.

Collaboration with Celebrities and Influencers: Space corporations collaborate with celebrities, technological know-how communicators, and social media influencers to promote planetary defense messages to their followers. Celebrities' involvement can appreciably expand the attain and impact of recognition campaigns.

Interactive Exhibits and Planetarium Shows: Museums, technological know-how centers, and planetariums characteristic interactive exhibits and planetarium shows associated with asteroids and impact activities. These handson reviews and immersive shows assist traffic grasp the magnitude of asteroid impact risks.

Community Engagement Events: Space companies and companies behavior network engagement activities, such as celebrity parties and skywatching occasions, to engage immediately with the general public. These events provide possibilities for human beings to find out about asteroids and planetary protection from experts.

A variety of conversation gear are used to growth public awareness approximately Potentially Hazardous Asteroids and planetary defense. From on-line structures and social media to instructional programs and network events, those tools play a crucial function in informing the public, encouraging preparedness, and fostering a sense of collective responsibility in safeguarding our planet from capability impact threats.

Education and outreach packages play a important role in growing public attention and understanding of Potentially Hazardous Asteroids (PHAs) and planetary protection. These applications purpose to interact people of every age and backgrounds, offering them with accurate statistics, fostering hobby in area technological know-how, and promoting preparedness for capacity asteroid effect occasions. Here are the important thing additives of training and outreach applications related to PHAs and planetary protection:

School Curriculum Integration: Space agencies and educational establishments collaborate to combine asteroidrelated subjects into college curricula. Lessons about asteroids, effect risks, and planetary protection are integrated into technology, astronomy, and earth science courses, making sure that scholars receive comprehensive know-how from an early age.

Educational Resources and Materials: Developing educational sources and materials is an critical component of outreach programs. These resources consist of truth sheets, brochures, interactive websites, movies, and academic video games which might be easily reachable to instructors, college students, and the overall public.

Classroom Presentations and Workshops: Experts in planetary defense and area technology behavior school room displays and workshops. These interactive periods allow college students to engage with specialists, ask questions, and deepen their know-how of asteroids and effect mitigation strategies.

Astronomy Clubs and Public Events: Astronomy golf equipment and technological know-how centers organize public activities, big name events, and observatory visits to introduce people to the wonders of the night time sky and train them about PHAs. Such events create possibilities for handson gaining knowledge of and encourage a fascination with space technology.

Public Lectures and Seminars: Universities, studies institutions, and area agencies arrange public lectures and seminars by professionals inside the area. These occasions are open to the general public and provide indepth insights into asteroid technology, planetary protection efforts, and the importance of space exploration.

Citizen Science Projects: Citizen technology tasks regarding the observation and reporting of asteroids are launched to actively contain the public in facts collection. Participants find out about asteroid identity, monitoring, and make contributions valuable records to professional astronomers.

Teacher Professional Development: Teacher expert development workshops are carried out to equip educators with the know-how and assets to efficiently educate about PHAs and planetary defense. Empowering instructors with correct facts ensures they are able to encourage and train their college students successfully.

Online Courses and Webinars: Online courses and webinars are designed to reach a wider target audience past conventional classroom settings. These courses cowl diverse aspects of asteroid science, impact dangers, and the modern studies in planetary protection.

Collaboration with Science Communicators: Space groups and organizations collaborate with technology communicators, influencers, and media shops to supply correct and engaging messages approximately asteroids and planetary defense. Collaboration with famous communicators increases the visibility and effect of outreach efforts.

International Collaboration and Global Initiatives: Education and outreach programs enlarge beyond countrywide borders thru international collaboration and global tasks. Coordinating efforts with different international locations guarantees a broader dissemination of knowledge and raises recognition on a international scale.

Education and outreach programs are instrumental in informing the public approximately Potentially Hazardous Asteroids and planetary defense. By leveraging numerous structures, attractive with instructional establishments, and regarding residents in medical endeavors, these packages foster a sense of duty and preparedness, empowering individuals to make a contribution to worldwide efforts in safeguarding our planet from potential asteroid impact threats.

7.3 International Cooperation and Policy Framework

Asteroids present one of the most significant existential threats to humanity. While the probability of an asteroid impact is low in any given year, the potential consequences of such an event—whether in the form of a mass extinction or severe global disruption—make it imperative for nations to work together in mitigating the risks posed by these celestial bodies. International cooperation and a cohesive policy framework are essential for addressing asteroid threats on a global scale.

As the scientific community has made clear, no single nation can address the challenge of asteroid collisions alone. The scale of the problem requires a united front, encompassing a variety of disciplines, from space agencies and scientific organizations to governments and the private sector. The importance of international cooperation in asteroid defense cannot be overstated, as the threat of a catastrophic asteroid impact is not confined to any one country. An asteroid impact, regardless of where it occurs, would have global ramifications. Therefore, cooperation among nations is not just beneficial but essential in preparing for and responding to this threat.

In the past, space exploration and planetary defense efforts were largely the domain of individual countries. However, as our understanding of the asteroid risk has evolved, it has become clear that a coordinated global effort is required to ensure the safety of humanity. Asteroids do not recognize national borders, and an impact from even a small asteroid could have far-reaching, global effects on ecosystems, economies, and societies. Therefore, international collaboration is crucial to maximize the effectiveness of detection, tracking, and mitigation efforts.

Several international agreements and partnerships have already been established to address asteroid threats. These frameworks provide the foundation for collective action, ensuring that all countries have access to asteroid data, are involved in global tracking efforts, and can contribute to the development of mitigation strategies. The space community, which includes space agencies, scientists, and engineers from various countries, is already working together to identify and track potentially hazardous asteroids (PHAs) and develop technologies that could potentially divert or mitigate impacts.

One of the most significant initiatives to foster international cooperation in the area of asteroid threat mitigation has been the involvement of the United Nations (UN). In 2001, the UN established the International Asteroid Warning Network (IAWN), which is a collaborative effort aimed at detecting and tracking near-Earth objects (NEOs). IAWN is comprised of a network of observatories, telescopes, and space agencies that share data on the locations and trajectories of asteroids. The goal of this network is to provide early warning of potential threats and give the global community time to develop countermeasures if necessary.

Alongside IAWN, the United Nations Office for Outer Space Affairs (UNOOSA) plays a key role in facilitating international discussions on planetary defense and fostering cooperation among countries to address the threat posed by asteroids. UNOOSA serves as a platform for the development of space-related policies and initiatives, bringing together spacefaring nations to address common challenges. As part of its efforts, the UN also organized the "Asteroid Day" initiative, an annual event held on June 30 to raise awareness about the asteroid threat and promote the development of global strategies for asteroid impact mitigation.

In addition to IAWN and UNOOSA, the UN has also advocated for the creation of an international asteroid mitigation policy framework that would ensure all countries are prepared for the eventuality of an asteroid impact. This framework would outline the steps that nations should take in the event of a potential asteroid collision, from early detection and impact modeling to implementing deflection or evacuation strategies.

• International Asteroid Warning Network (IAWN): A global effort to detect and track potentially hazardous asteroids.

• United Nations Office for Outer Space Affairs (UNOOSA): Facilitates international cooperation in space exploration and planetary defense.

• Asteroid Day: An annual event to raise awareness about asteroid impact risks and encourage international action.

In addition to the UN's role, space agencies around the world have taken the initiative to establish collaborations aimed at mitigating asteroid threats. The European Space Agency (ESA), NASA, the Russian space agency Roscosmos, the Indian Space Research Organisation (ISRO), the Chinese space agency CNSA, and other national space agencies are all involved in a variety of joint programs that involve sharing resources, data, and expertise on asteroid detection and impact risk assessment.

A prime example of international collaboration between space agencies is the AIDA (Asteroid Impact & Deflection Assessment) mission. This mission is a collaboration between NASA and the European Space Agency (ESA), which aims to test asteroid deflection techniques. Through the AIDA mission, NASA's Double Asteroid Redirection Test (DART) spacecraft will impact the moonlet of the binary asteroid system Didymos, and ESA's Hera spacecraft will follow to study the effects of the impact. The mission exemplifies how international partnerships can pool resources and expertise to tackle a common, shared threat.

Other international collaborations include joint asteroid missions such as NASA's OSIRIS-REx, the Japanese Hayabusa2, and China's future asteroid exploration missions. These joint missions share scientific data, contribute to planetary defense research, and foster cross-national cooperation in understanding and mitigating the asteroid risk.

• AIDA (NASA & ESA): Collaborative effort to test asteroid deflection techniques.

• OSIRIS-REx (NASA), Hayabusa2 (JAXA), and China's missions: Joint missions to study asteroids and improve planetary defense capabilities.

In addition to global cooperation, many countries have begun to implement national policies focused on planetary defense and asteroid impact mitigation. These policies are designed to ensure that governments have a coordinated and effective response in the event that an asteroid is determined to be on a collision course with Earth.

For example, the United States has developed a National Near-Earth Object Preparedness Strategy and Action Plan, which outlines steps for improving asteroid detection capabilities, developing deflection technologies, and establishing protocols for responding to potential asteroid threats. This document, created in collaboration with NASA and other federal agencies, provides a blueprint for US government actions related to planetary defense.

Similarly, other countries are beginning to develop their own national strategies to address asteroid impacts. In the European Union, the Space Situational Awareness Program (SSA) is focused on monitoring near-Earth objects and ensuring that European nations are prepared for potential impacts. China has also expressed interest in developing its own planetary defense program as part of its broader space exploration goals.

• US National Near-Earth Object Preparedness Strategy: A plan for improving asteroid detection, deflection, and impact mitigation in the United States.

• European Union's Space Situational Awareness Program: Focused on monitoring and tracking near-Earth objects.

• China's Planetary Defense Program: Part of China's broader space exploration and defense efforts.

In addition to governmental and intergovernmental efforts, the private sector also has a significant role to play in asteroid threat mitigation. Companies in the space industry, such as SpaceX, Blue Origin, and other private space ventures, are advancing the technologies needed for asteroid exploration and planetary defense. These companies could potentially provide valuable resources, technologies, and expertise to help mitigate asteroid threats in the future.

For example, private companies are working on developing technologies for asteroid mining, which could be used to extract resources from asteroids. While the primary goal of asteroid mining is to harness valuable materials, these technologies could also be adapted for planetary defense purposes, such as altering the trajectory of hazardous asteroids.

• SpaceX, Blue Origin, and other companies: Develop space technologies that could be applied to planetary defense.

• Asteroid Mining Technologies: Could be adapted for asteroid deflection and mitigation efforts.

International cooperation and a robust policy framework are essential for addressing the global threat posed by asteroid impacts. While the risk of an asteroid collision may seem distant, the potential consequences make it imperative for nations to work together to mitigate this risk. Through collaborative efforts by the United Nations, space agencies, national governments, and the private sector, humanity can develop the technologies and strategies necessary to detect, track, and ultimately deflect potentially hazardous asteroids. By strengthening international partnerships and developing cohesive policies, the global community can ensure that we are prepared for the possibility of an asteroid impact, safeguarding the future of life on Earth.

7.4. The Ethics of Withholding Threat Information from the Public

The query of whether to withhold records approximately potential international threats, consisting of close to-Earth object (NEO) influences or other catastrophic cosmic activities, from the general public touches on profound ethical, social, and political considerations. Balancing transparency with the obligation to save you panic, misinformation, and social disruption represents a undertaking confronted by using governments, scientific institutions, and global businesses. The ethical catch 22 situation revolves round the public's proper to realize versus the ability harms that untimely or uncontrolled disclosure might purpose. This anxiety increases fundamental questions on trust, obligation, autonomy, and the role of data in democratic societies.

At the middle of this debate is the precept of transparency. Democratic governance and medical integrity demand openness and duty. Citizens have the right to be knowledgeable about dangers that can have an effect on their protection and nicely-being. Transparency fosters believe among government and the public, encouraging cooperation and preparedness. When hazard statistics is shared openly, it empowers people and groups to make informed choices, take part in making plans and response efforts, and make contributions to collective resilience.

However, transparency is not an absolute price; it exists inside a framework of social responsibility. The communication of danger statistics incorporates capability risks, specifically whilst the information is uncertain or incomplete. Scientific predictions of impact possibilities frequently involve complicated calculations with margins of blunders, which can be misunderstood or sensationalized by way of the media and the general public. Premature announcements of high-danger eventualities that later show much less intense can erode credibility and purpose public confusion or complacency.

One moral argument for withholding certain hazard statistics is the prevention of panic and social disease. History offers examples where surprising disclosure of catastrophic news caused sizeable fear, monetary instability, or breakdowns in public order. In eventualities wherein an impact risk is distant or unsure, government may additionally choose to manipulate information flow carefully to avoid needless alarm. This approach presupposes a paternalistic model, in which professionals and leaders determine what statistics is suitable to percentage and while.

Yet, paternalism raises issues about autonomy and respect for individuals' capability to handle tough truths. Withholding information may be perceived as patronizing or manipulative, undermining agree with in institutions. The erosion of public self belief may be extra damaging than the preliminary worry as a result of disclosure. Moreover, individuals and communities frequently have the functionality to engage in significant instruction if competently informed. Ethical conversation techniques thus emphasize honesty balanced with contextualization and assist.

Another dimension of this moral issue concerns equitable get admission to to records. The selective dissemination of danger know-how risks developing disparities, in which positive groups or international locations are privileged with early warning while others stay uninformed. Such inequalities can exacerbate vulnerabilities and social tensions, tough ideas of justice and fairness. Global threats like asteroid impacts require inclusive and coordinated records sharing that respects the rights and desires of all affected populations.

The role of media and social networks complicates the ethical landscape. In the digital age, controlling records is increasingly more tough. Attempts to withhold or delay risk information may be circumvented via leaks or alternative channels, probably resulting in incorrect information, rumors, and distrust. This dynamic argues for proactive transparency and engagement with media stores to make certain accurate and responsible reporting. Ethical recommendations for journalists and communicators grow to be crucial to navigate the quality line between alerting the general public and avoiding sensationalism.

International cooperation provides layers of complexity and necessity. Planetary threats transcend country wide borders, requiring collective selection-making approximately information disclosure. Different international locations might also have varying hazard tolerances, political cultures, and verbal exchange guidelines. Establishing shared moral frameworks and protocols via global establishments helps harmonize strategies, build mutual trust, and coordinate public messaging. Transparency in such frameworks reinforces legitimacy and effectiveness.

Ethical considerations additionally increase to the timing of disclosure. The ability impact timeline influences whether instant public notification is prudent or if a phased technique is most popular. Early caution years or a long time in advance permits for measured practise and technological response improvement, whereas drawing close threats can also call for speedy, complete alerts coupled with emergency control. The ethical vital is to calibrate conversation techniques to the character and immediacy of the risk.

The concept of informed consent emerges in discussions approximately public knowledge of threats. While complete consent inside the conventional sense might not be attainable in mass communication contexts, supplying clear, trustworthy, and accessible data respects people' rights to be informed individuals in societal chance control. Educational initiatives about cosmic dangers make a contribution to an informed public capable of understanding and engaging with threat records responsibly.

Philosophical perspectives on the ethics of withholding danger data also recall the value of hope as opposed to worry. Overemphasis on catastrophic capacity without context or positive steerage can foster despair and fatalism, hindering adaptive behaviors. Ethical verbal exchange seeks to stability realism with encouragement, highlighting ongoing mitigation efforts, scientific progress, and possibilities for resilience.

In sensible phrases, the status quo of specialised conversation teams trained in crisis and danger communication is a vital ethical exercise. These groups tailor messages to diverse audiences, use clean and consistent language, and offer actionable recommendation. Incorporating remarks from social scientists, ethicists, and community leaders enhances the relevance and acceptance of hazard facts.

The ethics of withholding threat facts from the public entails navigating a complicated terrain where transparency, social responsibility, agree with, autonomy, equity, and effective communique intersect. While there can be justified reasons for careful statistics management, openness stays foundational to democratic values and public empowerment. Ethical frameworks and collaborative worldwide efforts are vital to ensure that statistics about worldwide threats is handled in ways that admire human dignity, foster preparedness, and guide collective protection.

CHAPTER 8

Future Threats and Precautions

8.1. Impact of Advances in Science and Technology

Artificial Intelligence (AI) and Machine Learning (ML) have revolutionized numerous fields, along with planetary protection and the observe of Potentially Hazardous Asteroids (PHAs). These advanced technologies offer powerful equipment to investigate information, make predictions, and enhance our expertise of asteroids' characteristics and impact risks. Here are some applications of AI and ML within the context of PHAs:

Asteroid Detection and Classification: AI and ML algorithms are used to analyze telescope pix and pick out capability asteroids. By automating the detection method, astronomers can efficiently locate new PHAs and NEOs, enhancing early warning competencies.

Orbit Prediction and Impact Risk Assessment: ML models can expect asteroid orbits more correctly by using analyzing historical facts and thinking about various gravitational influences. This allows better impact danger tests and lets in for the identification of future close tactics to Earth.

Deflection Mission Planning: AI and ML assist in planning deflection missions for PHAs that pose a enormous impact danger. By simulating exceptional deflection strategies and thinking about complex orbital dynamics, researchers can optimize challenge trajectories and maximize their success.

Data Analysis and Pattern Recognition: AI and ML algorithms can system enormous amounts of records from telescopes, radars, and spacecraft missions to identify styles and extract treasured insights about asteroids' composition, length distribution, and rotation traits.

Asteroid Shape and Spin Estimation: ML fashions can analyze light curves and radar statistics to estimate the form and spin parameters of asteroids. This facts is important for expertise an asteroid's rotational conduct and planning deflection missions.

Impact Consequence Modeling: AI and ML strategies can simulate the ability effect of an asteroid on Earth, predicting the possible effects in terms of blast outcomes, crater formation, and tsunamis. This information allows assess the effect's severity and manual emergency preparedness.

Spacecraft Autonomy and Navigation: AI and ML are hired in spacecraft autonomy to enhance navigation and optimize gas intake during deflection missions. Smart algorithms can enable self sufficient decisionmaking and path corrections in realtime.

Anomaly Detection and Risk Mitigation: ML algorithms can pick out ability facts anomalies or uncertainties in asteroid tracking, making sure that capability impact dangers aren't unnoticed and supplying more sturdy risk mitigation strategies.

Data Fusion and MultiSensor Integration: AI strategies are used to fuse statistics from distinctive sensors, such as optical telescopes, radar structures, and infrared instruments, to create a complete picture of an asteroid's characteristics and conduct.

RealTime Threat Assessment: AI and ML systems can constantly analyze incoming information to provide realtime threat evaluation, permitting speedy responses and coordination in case of highimpact risk scenarios.

The packages of Artificial Intelligence and Machine Learning in planetary defense and the take a look at of Potentially Hazardous Asteroids are various and powerful. By leveraging these technology, researchers can decorate detection, prediction, and mitigation techniques, ultimately enhancing our capacity to protect our planet from ability asteroid impact threats. As AI and ML maintain to boost, their function in planetary defense will probably grow to be even extra full-size in the future.

Automatic observation and collision prediction systems are crucial components of planetary protection efforts geared toward detecting, tracking, and predicting the trajectories of Potentially Hazardous Asteroids (PHAs) and other NearEarth Objects (NEOs). These systems leverage advanced technologies, together with Artificial Intelligence (AI) and Machine Learning (ML), to automate the manner of monitoring celestial gadgets and identifying capability effect risks. Here are the key factors of automatic observation and collision prediction:

Automated Telescope Networks: Sophisticated telescope networks geared up with robot manage systems enable non-stop and automatic observations of the night time sky. These telescopes test predefined regions of area to come across and tune moving objects, including asteroids.

Image Analysis and Object Identification: AI and ML algorithms examine telescope photos to become aware of and classify asteroids among the widespread quantity of detected celestial objects. Realtime image analysis lets in for the speedy identification of recent PHAs and NEOs.

Orbital Prediction and Trajectory Analysis: Using historic and realtime commentary statistics, AIbased algorithms are expecting the orbits of asteroids. Trajectory analysis takes under consideration gravitational forces and perturbations from planets and different celestial our bodies to assess capability impact risks.

Collision Probability Calculation: By combining orbit predictions with Earth's orbital parameters, AI systems calculate the collision opportunity of detected asteroids with our planet. This lets in for the identity of asteroids that pose a good sized impact chance.

Impact Risk Assessment and Ranking: Automatic systems check the severity of ability affects based totally on asteroid length, pace, and the opportunity of collision. The ranking of PHAs helps prioritize resources and attention on those with higher impact risks.

Early Warning and Alert Systems: Automatic observation and collision prediction systems provide early warning signals when a brand new PHA is detected and recognized as a potential impact danger. Rapid notification enables well timed reaction and in addition monitoring.

Anomaly Detection and Data Quality Control: AI and ML algorithms assist in anomaly detection to perceive mistakes or uncertainties in observation statistics. Data exceptional manipulate measures ensure the accuracy and reliability of predictions and impact chance assessments.

Continual Monitoring and Data Fusion: Automated systems usually display PHAs and NEOs to tune any modifications in their orbits or behavior. Data fusion techniques integrate statistics from numerous sources, inclusive of telescopes and radar, to improve accuracy.

RealTime Collision Prediction and DecisionMaking: In highimpact risk eventualities, computerized systems provide realtime collision predictions. These predictions guide decisionmakers in formulating suitable deflection or mitigation strategies.

Continuous Improvement and Learning: AI and ML structures study from new information and observations, constantly enhancing their overall performance and accuracy. As greater data will become available, the systems refine their fashions and predictions.

Automatic statement and collision prediction systems are crucial for early detection and assessment of Potentially Hazardous Asteroids and NEOs. Leveraging superior technology like AI and ML, those systems enable realtime tracking, correct orbital predictions, and effect chance tests, ensuring green and effective planetary defense efforts. The continuous development of these systems through gaining knowledge of and facts refinement enhances our potential to safeguard our planet from capacity asteroid impact threats.

8.2. ForwardLooking Predictions and Scenarios

Risk mitigation is a vital component of planetary protection efforts, especially concerning Potentially Hazardous Asteroids (PHAs) and NearEarth Objects (NEOs). As we explore area and do not forget the future of human presence beyond Earth, know-how and addressing the risks posed by PHAs become more and more important. Here are the key concerns regarding risk mitigation and the future of human presence in area:

Deflection and Mitigation Strategies: To deal with the impact dangers of PHAs, diverse deflection and mitigation techniques are being explored. These techniques include kinetic impactors, gravity tractors, nuclear deflection, and laser ablation, amongst others. Research and testing are ongoing to determine the handiest and safe strategies for altering the trajectories of potentially dangerous asteroids.

International Collaboration and Coordination: Planetary defense calls for global collaboration and coordination amongst area businesses, governments, and global companies. Sharing statistics, understanding, and sources is vital for a unified reaction to ability impact threats.

Planetary Defense Exercises and Drills: Regular sporting activities and drills are performed to simulate asteroid impact eventualities and check response plans. These sporting activities contain relevant stakeholders to beautify preparedness and make certain a coordinated method in case of a real impact danger.

Public Awareness and Education: Raising public awareness about PHAs and planetary defense is crucial. Educating the general public approximately the risks, effect consequences, and mitigation efforts fosters assist for area missions and advocacy for expanded investment on this field.

Monitoring and Tracking Systems: Continuous monitoring and tracking of PHAs are crucial to hit upon any modifications of their trajectories. Advanced telescopes, radar systems, and area missions provide valuable statistics for greater correct predictions and danger checks.

Human Space Exploration and Asteroid Research: As humanity ventures deeper into space, understanding asteroids becomes important for human space exploration protection.

Studying asteroids up near, including through sample return missions, contributes to our understanding of their houses and conduct.

Space Infrastructure Resilience: Future human missions to other celestial bodies, just like the Moon or Mars, will require strong area infrastructure to ensure the safety of astronauts. Preparing for capacity asteroid threats to area habitats and bases is an vital consideration.

Planetary Defense Legislation and Policies: Governments and worldwide groups are working on establishing planetary protection legislation and policies. These frameworks address issues related to asteroid discovery, impact risk assessment, and coordination for capability mitigation missions.

Development of Space Technologies: Advancements in space technologies, such as propulsion systems, spacecraft autonomy, and robotics, help extra effective deflection missions and beautify our capability to respond to potential effect dangers.

Space Resource Utilization: Developing area resource usage techniques can guide planetary defense efforts. Insitu useful resource usage on asteroids could provide the essential substances for deflection missions or infrastructure production.

Risk mitigation is a essential component of planetary protection, and efforts on this field are crucial as humanity ventures in addition into area. Collaboration, tracking structures, advanced technology, public recognition, and a focal point on human space exploration protection all play a pivotal position in making sure our ability to address capability impact risks from Potentially Hazardous Asteroids. By proactively making ready and advancing our knowledge and competencies, we can protect our planet and destiny human presence in space.

Managing capacity threats from Potentially Hazardous Asteroids (PHAs) requires a multifaceted method that entails early detection, correct impact danger assessment, and proactive hazard mitigation strategies. Here are the important thing factors of dealing with capacity asteroid threats:

Early Detection and Monitoring: Continuous monitoring of the night sky the usage of superior telescopes and radar systems is crucial for early detection of PHAs. Early warning lets in for adequate time to evaluate the asteroid's trajectory and potential effect dangers.

Impact Risk Assessment: Sophisticated computational fashions and simulations are used to assess the impact risks posed by way of PHAs. Factors inclusive of length, pace, and the possibility of collision are taken into consideration to decide the severity of capacity impact occasions.

Deflection Missions: In the occasion of a PHA with a good sized impact risk, deflection missions can be taken into consideration. These missions involve sending spacecraft to the asteroid to alter its trajectory and divert it far from Earth.

Research and testing of various deflection techniques are ongoing.

Emergency Preparedness and Response Plans: Governments and relevant government expand emergency preparedness and response plans to address capability effect situations. These plans consist of evacuation strategies, communication techniques, and coordination among different organizations.

International Collaboration: Planetary protection is a worldwide enterprise, and worldwide collaboration is vital for sharing data, knowledge, and sources. Collaborative efforts allow a more comprehensive method to dealing with ability threats from PHAs.

Research and Observation: Continuous research and remark of PHAs are critical for improving effect risk checks and deflection techniques. Understanding the traits and behavior of asteroids provides precious insights into ability threat management.

Public Awareness and Education: Public awareness campaigns and educational programs tell humans about PHAs and planetary defense efforts. Raising cognizance allows garner public assist, encourages advocacy for expanded investment, and fosters a experience of collective duty.

Planetary Defense Exercise Drills: Conducting regular physical games and drills to simulate ability impact scenarios facilitates put together relevant stakeholders for effective responses. These sporting events check reaction plans and become aware of areas for development.

Space Traffic Management: As the quantity of satellites and spacecraft in space will increase, space traffic control turns into important for keeping off collisions with PHAs and different area particles. Cooperation among space corporations and business entities is critical to maintain safe orbital environments.

Continual Improvement and Adaptation: The area of planetary defense is constantly evolving, and classes found out from every encounter with PHAs contribute to refining reaction strategies. Flexibility and flexibility are critical to live ahead of capacity threats.

Managing potential threats from Potentially Hazardous Asteroids requires a comprehensive method that encompasses early detection, correct effect threat evaluation, and proactive danger mitigation techniques. Through worldwide collaboration, public focus, continuous studies, and emergency preparedness, we will higher safeguard our planet and destiny human presence in space from capability effect activities.

8.3 The Role of AI and Machine Learning in Asteroid Detection

In the pursuit of safeguarding Earth from the potential threat posed by asteroids, new technological advancements are pivotal in enhancing our ability to detect and track Near-Earth

Objects (NEOs). While traditional methods of asteroid detection, such as optical telescopes and radar systems, have been incredibly valuable, they face limitations in terms of speed, accuracy, and the ability to process vast amounts of data in real-time. To overcome these challenges and better prepare for the future, Artificial Intelligence (AI) and Machine Learning (ML) are increasingly being integrated into asteroid detection systems, revolutionizing how we identify, monitor, and assess the risks associated with NEOs.

AI and ML technologies can significantly enhance the effectiveness of existing asteroid detection techniques by offering the capability to rapidly analyze enormous amounts of data, improve precision in identifying potential threats, and ultimately increase our chances of early detection.

The sheer volume of data generated by asteroid surveys, satellite missions, and space telescopes is growing exponentially. For example, current surveys such as the Pan-STARRS and the Catalina Sky Survey capture millions of images, each containing thousands of data points related to the position, movement, and trajectory of asteroids. With so much data being collected on a daily basis, manual analysis becomes a daunting task, one that is slow and prone to human error. This is where AI and ML step in.

Machine learning, a subset of AI, allows computers to learn from large datasets without being explicitly programmed to identify patterns. In asteroid detection, machine learning algorithms can be trained to detect subtle changes in the movement or position of asteroids in real-time, processing enormous amounts of data far faster than human analysts could. Through deep learning and neural networks, these systems can improve their predictive accuracy over time as they are exposed to more data, enabling better predictions about the orbits and future trajectories of asteroids.

AI-powered systems also offer the potential for more precise classification of asteroids. While current methods of classification are based primarily on visual observation and simple categorizations, machine learning algorithms can analyze complex features such as the chemical composition, size, and rotation patterns of an asteroid, thus refining our understanding of these celestial objects and their potential threat levels.

AI and ML technologies offer a variety of tools and approaches that significantly enhance asteroid detection capabilities, both in terms of speed and accuracy.

1. Data Processing and Analysis: AI and ML can handle large-scale data sets from multiple sources, including optical telescopes, radar, and infrared surveys, at a speed and efficiency that would be impossible for human researchers to match. Algorithms can automatically sift through this data, flagging potential asteroid candidates that might otherwise go unnoticed. Additionally, ML algorithms can be used to detect anomalies or unexpected patterns in the data that suggest the

presence of previously undetected asteroids or asteroids on potential collision courses.

For instance, NASA's Asteroid Terrestrial-impact Last Alert System (ATLAS) is already using AI to process images and detect asteroids that may be on a collision course with Earth. The system utilizes ML algorithms to compare current images with historical data, enabling rapid identification of new or potentially hazardous asteroids.

2. Improved Orbit Prediction: Machine learning can be used to more accurately predict the orbits of asteroids. By analyzing the gravitational influences and past orbital data, ML algorithms can generate more precise models for predicting asteroid paths over long periods of time. This allows for better forecasting of potential collisions with Earth, even decades or centuries in advance. AI models can also assess the impact of different variables, such as gravitational interactions with other celestial bodies, on an asteroid's trajectory, giving us a more complete picture of future threats.

3. Automating Asteroid Classification: Machine learning algorithms are particularly well-suited for classifying asteroids based on a variety of physical characteristics. Asteroids come in a wide range of shapes, sizes, and compositions, and accurately categorizing them is essential for determining their potential impact risks. Using ML models, asteroid data—such as their brightness, shape, size, and spectrum—can be classified

automatically, streamlining the process of identifying which asteroids pose the greatest threat to Earth.

One key area where AI could revolutionize asteroid classification is in determining whether an asteroid is a "potentially hazardous asteroid" (PHA). PHAs are defined by their proximity to Earth and their size, and AI models can help identify these objects more efficiently. By analyzing an asteroid's characteristics, AI can assign it a level of risk based on its potential to impact Earth, providing valuable information for mitigation planning.

4. Real-Time Monitoring and Alert Systems: Real time monitoring is another area where AI and ML have the potential to significantly improve asteroid detection. Traditional detection methods often rely on scheduled observations, but AI-powered systems can continuously analyze data and update asteroid tracking information in real-time. This allows for faster response times when a potentially hazardous asteroid is detected, reducing the time between discovery and action.

AI systems can also play a crucial role in early warning systems. As AI algorithms become more advanced, they can analyze real-time data streams from observatories and telescopes, issuing immediate alerts when an asteroid is detected that might threaten Earth. This rapid response is essential for implementing mitigation strategies, such as asteroid deflection or evacuation plans, should the need arise.

The integration of AI and machine learning into current and upcoming space missions is critical for enhancing asteroid detection capabilities. Many space missions, such as NASA's OSIRIS-REx, JAXA's Hayabusa2, and ESA's Hera mission, aim to study asteroids and collect vital data that will inform planetary defense strategies. AI-powered systems onboard these spacecraft can process data in real-time, identifying asteroids, tracking their movements, and analyzing their characteristics autonomously.

For example, AI algorithms could be used to process the imagery collected by these missions, rapidly identifying asteroid features such as surface textures, composition, and any changes in the asteroid's rotation. By analyzing this data on-board, AI could help to refine our understanding of asteroid behavior and provide essential information for devising mitigation strategies in the future.

Beyond detection, AI also holds promise for improving asteroid deflection strategies. Once a potentially hazardous asteroid is detected, deflection is one of the most viable methods of preventing an impact. AI can optimize deflection strategies by modeling various impact scenarios and testing different deflection techniques, such as kinetic impactors or gravity tractors, to determine the most effective approach.

Machine learning algorithms can simulate numerous deflection scenarios, taking into account the size, speed, and composition of the asteroid, as well as the technology available for mitigation. By running these simulations, AI can help develop the most efficient strategies for altering an asteroid's course, allowing scientists and space agencies to prepare a set of responses in advance of a potential collision.

As AI and machine learning technologies continue to evolve, their role in asteroid detection will become even more critical. The potential for AI to revolutionize planetary defense is vast, and ongoing research and development will likely yield even more sophisticated systems for detecting, tracking, and mitigating asteroid threats.

The integration of AI with space telescopes, satellite networks, and asteroid missions will create a more comprehensive and robust asteroid defense system. In the future, AI could even be used to develop automated asteroid mining systems, which would not only provide valuable resources but could also be repurposed for planetary defense purposes, such as altering the course of hazardous asteroids.

As we look to the future, it is clear that AI will play a central role in safeguarding humanity from asteroid threats. By combining the capabilities of AI and machine learning with global detection networks and space exploration missions, we will be better equipped to identify potential asteroid impacts and take proactive measures to prevent disaster. In doing so, AI will help ensure the continued safety and survival of life on Earth.

8.4. Planetary Defense within the Context of Global Security

The protection of Earth from extraterrestrial threats, extensively the ones posed by close to-Earth gadgets (NEOs) which includes asteroids and comets, has increasingly more end up identified now not handiest as a systematic and technical challenge however as a count number of world protection. Planetary protection—the coordinated attempt to locate, display, and mitigate threats originating from space—intersects with the broader area of global security in profound and complicated ways. This convergence elevates planetary protection from a niche clinical pursuit into a strategic priority requiring the combination of medical knowledge, defense coverage, global cooperation, and emergency preparedness.

Global security historically makes a speciality of troubles consisting of geopolitical stability, military deterrence, cybersecurity, terrorism, and the proliferation of weapons of mass destruction. The addition of planetary defense expands the security schedule beyond terrestrial borders and conventional domain names. Unlike conventional threats that stand up from human actors or states, planetary hazards are herbal phenomena—accidental, indiscriminate, and capable of inflicting devastation on scales starting from local catastrophes to mass extinction activities. This fundamental difference demanding situations present protection frameworks and demands revolutionary techniques that go beyond political rivalries and navy posturing.

One key measurement of planetary defense in worldwide security is the recognition that asteroid effect dangers are inherently transnational. A large impact event could affect more than one countries concurrently, rendering countrywide responses insufficient. No single country possesses the sources or technological abilities to address a considerable danger by myself, necessitating strong worldwide collaboration. This has spurred the development of multilateral agencies and treaties aimed at fostering cooperation in detection, records sharing, and reaction planning. The United Nations Office for Outer Space Affairs (UNOOSA), for instance, performs a pivotal function in facilitating speak and coordinating efforts through frameworks just like the International Asteroid Warning Network (IAWN) and the Space Mission Planning Advisory Group (SMPAG).

Integrating planetary protection into international security additionally involves bridging the gap between civilian medical communities and navy institutions. While area businesses and astronomers lead the efforts in detection and research, defense ministries are an increasing number of engaged because of the results for national and worldwide safety. Military infrastructure, inclusive of radar structures and surveillance satellites, can enhance early caution talents. Conversely, protection information in crisis management, command and control, and logistics is crucial for orchestrating mitigation missions and civil protection measures. This dual-use capacity fosters new collaborations however additionally increases worries approximately weaponization and the militarization of area.

The strategic importance of planetary defense has catalyzed coverage discussions about governance, resource allocation, and the delineation of duties. Governments face the challenge of integrating planetary defense within current security doctrines whilst making sure transparency and averting duplication or conflict among corporations. National security strategies need to stability the prioritization of cosmic dangers in opposition to greater on the spot threats, weighing charges and blessings in budgeting and making plans. The involvement of private area actors, inclusive of business aerospace corporations, in addition complicates governance but also gives possibilities for innovation and performance.

From a security attitude, the development and deployment of planetary protection technologies—which include kinetic impactors, nuclear deflection gadgets, or gravity tractors—have dual-use implications that must be carefully managed. The ability militarization of space through the deployment of such technologies requires stringent controls and confidence-building measures to save you misunderstandings and escalation. International agreements and verification mechanisms are essential to make sure that planetary defense tasks stay peaceful and contribute to global balance rather than fueling hands races.

Emergency preparedness for effect scenarios integrates planetary protection with disaster management and humanitarian security. The prospect of a confirmed effect necessitates coordinated making plans among countrywide governments, global agencies, and civil society. This includes evacuation protocols, infrastructure resilience, communication strategies, and medical reaction. Lessons found out from herbal failures and pandemics inform those efforts, emphasizing the want for adaptable, inclusive, and scalable response frameworks. The involvement of security companies ensures that law enforcement, order renovation, and crisis management are incorporated into planetary defense contingencies.

Public verbal exchange and hazard notion are vital additives of planetary defense's protection measurement. Misinformation, panic, and mistrust can undermine both scientific and safety efforts. Transparent, consistent, and culturally sensitive communique strategies help construct public resilience and aid for mitigation measures. Security companies may deal with threats to facts integrity, such as disinformation campaigns or cyberattacks focused on planetary defense infrastructure. Safeguarding records and maintaining public confidence hence constitute important safety obligations.

International equity and justice shape ethical pillars inside the planetary defense-protection nexus. The unequal distribution of technological abilities and resources among countries raises questions on get right of entry to to records, participation in selection-making, and the sharing of protecting measures. Ensuring that developing international locations and inclined populations are blanketed in planetary protection tasks aligns with broader goals of worldwide security and sustainable development. Capacity building, era switch, and monetary aid mechanisms contribute to a greater just and powerful global defense posture.

The evolving chance panorama, which includes the possibility of unknown cosmic hazards or combined natural and human-made crises, demands adaptive safety tactics. Integrating planetary protection into global safety encourages interdisciplinary studies, scenario planning, and flexible governance. It additionally fosters resilience by way of linking planetary protection with weather security, pandemic preparedness, and area sustainability efforts, spotting the interconnectedness of world dangers.

Planetary protection represents an important frontier in international security, requiring the convergence of technology, coverage, navy know-how, and worldwide cooperation. Its transnational nature, technological complexity, and humanitarian stakes mission traditional safety paradigms and get in touch with for modern, inclusive, and collaborative frameworks. As humanity's awareness and competencies enhance, embedding planetary defense within the broader safety architecture may be imperative for safeguarding Earth's destiny within the cosmic environment.

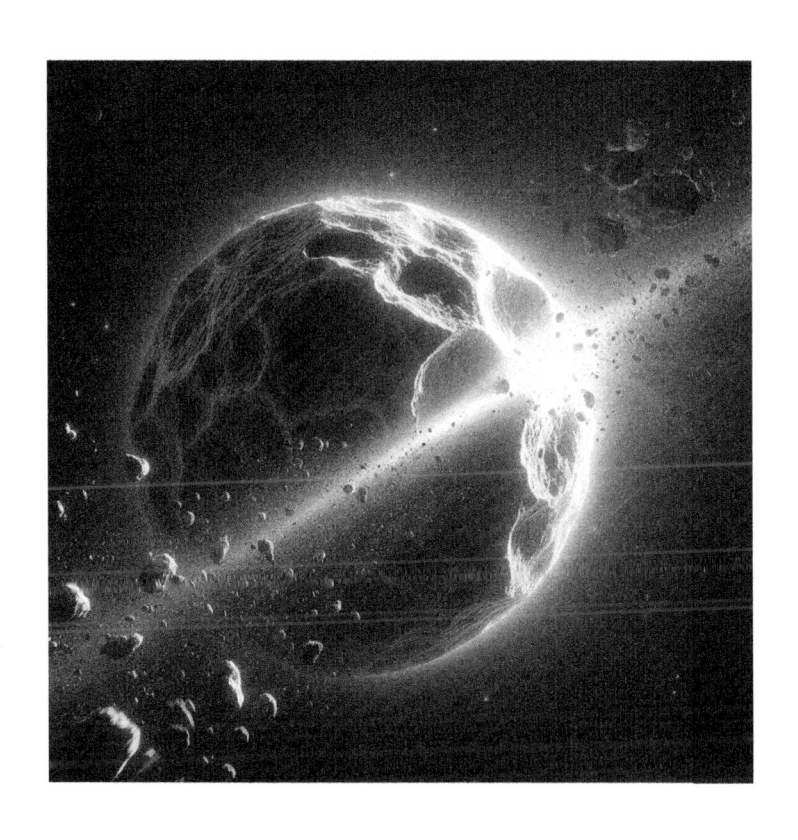

CHAPTER 9

Asteroid Mining and Ethical Questions

9.1. Commercial Interests vs Planetary Protection

The dawn of asteroid mining heralds a transformative technology in humanity's dating with space. What as soon as belonged completely to the area of technological know-how fiction is hastily becoming a tangible commercial enterprise, pushed by advances in era, developing call for for uncommon and valuable substances, and the promise of considerable financial profits. Companies and nations are eyeing the good sized mineral wealth contained in near-Earth asteroids as a potential technique to terrestrial useful resource scarcity. However, this industrial rush into the final frontier raises profound moral questions, specifically regarding planetary safety — the imperative to maintain both Earth and other celestial our bodies from dangerous infection or unintended consequences.

Asteroid mining offers get admission to to resources such as platinum organization metals, uncommon earth elements, water ice, and other substances crucial to modern generation and space exploration. These assets could fuel space infrastructure, support long-time period missions, and relieve pressures on Earth's finite reserves. Private organisations, which include startups and installed aerospace companies, are investing in prospecting missions, generation development, and felony frameworks to say and take advantage of asteroid sources. The pleasure surrounding this potential boom has spurred policy debates and induced calls for regulatory oversight to control competing interests and shield the distance surroundings.

Central to the ethical dialogue is the balance among business hobbies and planetary safety. On one facet, proponents argue that asteroid mining can boost up medical discovery, force monetary increase, and catalyze humanity's enlargement into the sun gadget. On the other, critics caution against the risks of environmental degradation in area, irreversible modifications to pristine celestial our bodies, and viable threats to Earth's biosphere thru contamination or particles generated by way of mining operations.

Planetary safety encompasses important worries: forward contamination, which entails introducing Earth-based microbes or substances to other celestial our bodies, doubtlessly disrupting their environments and medical investigations; and backward contamination, the opposite risk of bringing extraterrestrial organisms or substances lower back to Earth, which may pose organic hazards. Although the opportunity of existence on maximum mined asteroids is taken into consideration low, these concerns can't be brushed off, mainly as missions grow to be greater complex and the volume of cloth transported will increase.

The felony and regulatory landscape adds complexity to this moral dilemma. The Outer Space Treaty of 1967, which remains the foundational framework for space activities, prohibits dangerous contamination and country wide appropriation of celestial our bodies however leaves ambiguity regarding commercial exploitation. More latest treaties and tips, inclusive of the Moon Agreement, have restrained adoption and do now not comprehensively address asteroid mining. Consequently, countries have begun enacting countrywide laws to furnish mining rights, developing a patchwork regulatory environment that raises questions about fairness, jurisdiction, and enforcement.

Transparency and worldwide cooperation are critical moral pillars in navigating industrial and planetary protection tensions. Mining operations hazard creating conflicts over property rights, environmental requirements, and get right of entry to to medical facts. Ensuring that activities are carried out responsibly calls for strong oversight mechanisms, environmental impact checks, and mechanisms for dispute resolution. The involvement of international bodies, which includes the United Nations Committee on the Peaceful Uses of Outer Space (COPUOS), is critical for fostering harmonized guidelines that balance economic development with protection.

The environmental ethics of space mining additionally extend to the stewardship of celestial our bodies as a part of humanity's shared heritage. Many advocate for a precautionary precept, urging that mining sports be limited or carefully managed to keep away from irreversible alteration of scientifically treasured or culturally big sites. This perspective aligns with growing calls to view outer area no longer simply as a useful resource reservoir however as a domain warranting respect and conservation, much like Earth's own herbal ecosystems.

Furthermore, the capability influences on Earth's environment and society demand moral scrutiny. Mining operations could generate space particles, increasing collision dangers in already congested orbital paths. The extraction and transportation of substances can also consume considerable electricity and produce carbon emissions, raising questions about sustainability and weather effect. Moreover, the chance of sizeable income concentrated amongst some entities may want to exacerbate financial inequalities and geopolitical tensions, underscoring the want for equitable gain-sharing preparations.

In addition to environmental and criminal concerns, moral issues arise from the potential disruption of scientific studies. Asteroids serve as time tablets preserving primordial fabric which could unencumber secrets and techniques of the sun device's formation and evolution. Mining activities hazard contaminating or destroying these invaluable medical facts. Balancing industrial exploitation with clinical maintenance requires talk and collaboration among stakeholders to ensure that mining does not compromise humanity's collective information.

Public engagement and ethical reflection are crucial as humanity ventures into asteroid mining. Diverse perspectives, which includes indigenous worldviews, cultural values, and philosophical concerns approximately humanity's function inside the cosmos, increase the discourse and assist form accountable guidelines. Ethical frameworks need to combine social justice, intergenerational obligation, and recognize for the intrinsic cost of celestial environments.

The intersection of commercial pastimes and planetary safety in asteroid mining provides a multifaceted ethical assignment. While the economic and technological capacity is large, it have to be pursued with caution, duty, and a dedication to maintaining the space surroundings and Earth's safety. Developing comprehensive regulatory regimes, fostering worldwide collaboration, and tasty with diverse ethical views are vital steps closer to ensuring that asteroid mining advantages humanity with out compromising planetary stewardship.

9.2. Legal and Political Ramifications of Resource Claims

The prospect of extracting precious assets from asteroids and different celestial bodies introduces complicated felony and political challenges that increase some distance past the technical and economic dimensions of area mining. As humanity transitions from exploration to exploitation of outer area assets, the question of who owns those assets, under what situations they may be claimed, and how conflicts are resolved will become essential. These problems implicate foundational ideas of international law, sovereignty, countrywide pursuits, geopolitical power dynamics, and emerging area governance frameworks.

At the core of these legal and political debates lies the tension among the principle of the "not unusual heritage of mankind" and the dreams of character international locations and personal entities to steady specific rights to space sources. The Outer Space Treaty (OST) of 1967, the primary prison instrument regulating sports in space, establishes that outer space, inclusive of the Moon and other celestial our bodies, is not situation to countrywide appropriation by declare of sovereignty, use, profession, or another means. Article II explicitly prohibits states from claiming territorial sovereignty over extraterrestrial territories. This precept ambitions to save you the extension of terrestrial colonialism into space and preserve space for peaceful and equitable use.

However, the OST's silence on personal property rights and useful resource extraction leaves ambiguity. While the treaty bans sovereignty claims, it does no longer explicitly forbid the appropriation of resources extracted from celestial our bodies. This ambiguity has allowed some international locations, along with the USA and Luxembourg, to bypass country wide rules spotting non-public belongings rights over sources mined in space, efficiently granting prison truth to their commercial area industries. The U.S. Commercial Space Launch Competitiveness Act (2015) and Luxembourg's Space Resources Law (2017) exemplify this technique, permitting businesses to personal, use, and promote space resources, whilst putting forward compliance with worldwide duties.

These unilateral legislative moves have sparked excessive debate approximately their compatibility with the OST and their effect at the worldwide area governance regime. Critics argue that countrywide laws recognizing resource possession danger undermining the treaty's non-appropriation precept and will encourage a "space rush," main to conflicts over claims and get right of entry to. They warn of a fragmented prison landscape wherein effective nations and organizations would possibly dominate useful resource extraction, sidelining much less-developed international locations and global pastimes. The ability for disputes over claim obstacles, overlapping rights, and environmental responsibilities also raises concerns about felony uncertainty and enforcement challenges.

Political ramifications of space useful resource claims are profound. The strategic price of space sources, particularly water ice for gas manufacturing and uncommon metals for enterprise, elevates space mining to an arena of geopolitical competition. Nations understand manipulate over those sources as crucial to securing their destiny economic competitiveness, technological management, and space exploration skills. This dynamic may want to accentuate rivalries and initiate militarization of space if resource claims emerge as entangled with country wide protection priorities.

International efforts to address these challenges have resulted in multilateral projects aimed at developing norms, tips, and cooperative mechanisms. The United Nations Committee on the Peaceful Uses of Outer Space (COPUOS) has been a forum for speak, emphasizing the importance of transparency, advantage-sharing, environmental protection, and peaceful use. The Hague International Space Resources Governance Working Group, launched in 2017, brought together stakeholders from governments, industry, academia, and civil society to discover criminal and regulatory alternatives. These projects are seeking to balance commercial pursuits with the preservation of area as a global commons.

A key political query issues the status quo of an international framework for licensing, monitoring, and adjudicating space useful resource activities. Unlike terrestrial mining, which operates underneath well-advanced country wide and worldwide legal guidelines, space mining lacks a comprehensive regulatory regime. Proposals variety from developing an global area resources authority to making use of current bodies, together with the International Telecommunication Union or the International Seabed Authority's model for ocean mining, to space governance. Such frameworks ought to provide dispute resolution mechanisms, environmental oversight, and equitable get admission to provisions.

The prospect of aid claims additionally intersects with problems of sovereignty and jurisdiction in space. While the OST prohibits sovereignty over celestial our bodies, sports carried out in space must nonetheless be authorized and supervised through the suitable state birthday party, which stays responsible underneath international law. This introduces questions about how countrywide jurisdictions expand to private actors running in area, how to coordinate a couple of states' responsibilities, and a way to prevent conflicts between entities from unique international locations engaged in useful resource extraction.

Moreover, political worries consist of the threat of "area colonialism," wherein wealthier countries and agencies exploit area assets with restricted blessings to developing countries or the international network. Addressing this inequity calls for integrating concepts of justice and inclusion into area coverage, such as mechanisms for gain-sharing, era transfer, and capability-building. The United Nations' concept of the "not unusual historical past of mankind" envisions space assets as a shared asset, calling for global cooperation to make certain that exploitation advantages all humanity in preference to a privileged few.

Environmental considerations are increasingly shaping criminal and political discussions. Space mining has the capability to cause bodily alterations to celestial our bodies, generate debris, and intervene with scientific research and exploration. Legal frameworks must therefore include environmental safeguards, impact exams, and remediation responsibilities to limit damage and hold area for future generations. These concerns echo terrestrial environmental law but must be adapted to the precise situations of the gap surroundings.

Technological advances will likely boost up the tempo of area resource claims, intensifying prison and political challenges. Emerging skills in autonomous spacecraft, in-situ aid utilization (ISRU), and industrial area transport increase the feasibility and beauty of mining operations. This makes pressing the establishment of clear, constant, and honest guidelines governing space resource activities before conflicts get up.

The legal and political ramifications of useful resource claims in space gift a complicated matrix of demanding situations that require careful negotiation and worldwide collaboration. Balancing countrywide objectives, private company, global law, environmental protection, and equitable access is important to avoid battle and make sure the sustainable improvement of space sources. As humanity stands on the brink of a brand new frontier, the selections made these days regarding governance and international relations will shape the trajectory of area exploration and exploitation for generations to return.

CHAPTER 10

Conclusion

10.1. Summary and Evaluation

Asteroid threats are ability risks lurking in the depths of area, and they are able to have critical outcomes for our planet. Potentially Hazardous Asteroids (PHAs) are celestial bodies whose orbits intersect or closely technique Earth's orbit. These PHAs vary in length, starting from a few meters to thousands of kilometers in diameter, and they circulate on orbits that could doubtlessly lead to risky collisions.

The potential of asteroids colliding with Earth is taken into consideration a big threat to humanity and the Earth's environment. Throughout history, asteroid impacts have induced essential disasters. The impact effects can encompass massive craters on the collision website, surprise waves, tsunamis, fires, and even a nuclear winterlike impact. Therefore, tracking and managing asteroid threats have come to be a vital precedence for area scientists, astronomers, and researchers.

One of the foremost challenges of asteroid threats is early detection. The vastness of space round Earth makes it hard to spot those ability risks, and often, PHAs can only be identified while they come close to Earth or pass via our planet. As a end result, superior observatories, radar structures, and space missions play a crucial function inside the efforts to detect capacity threats.

Assessing potential threats and undertaking danger analysis involves knowledge the motion of asteroids and predicting their potential effect risks. Advanced laptop modeling and synthetic intelligence algorithms are applied to expect asteroid orbits, velocities, and viable collision factors. These analyses bear in mind elements which includes collision opportunity, asteroid diameter, and kinetic energy to determine the ability danger degree of a specific asteroid.

Various methods had been proposed to address asteroid threats. Strategies encompass space missions to modify the course of asteroids, nuclear explosions for redirection, highenergy lasers for ablation, and the usage of spacebased observatories and radar systems to stumble on and song asteroids with ability collision chances. These strategies are at the leading edge of clinical and technological efforts to address asteroid threats.

Despite those efforts, we nonetheless do no longer have a complete understanding of asteroid threats, and some capability future threats can also stay undetected. Therefore, continuous observation, studies, and attention of how to reply to a potential collision occasion are vital.

Asteroid threats are considered a extensive chance to humanity and our planet's ecosystem. Managing those threats and minimizing ability risks from risky celestial our bodies require continuous commentary, evaluation, and technological advancement. Space exploration and medical collaboration play a key function in better preparedness and growing a safer destiny for Earth.

Precautionary and preparedness measures are crucial components of addressing capability threats posed by way of Potentially Hazardous Asteroids (PHAs) and other NearEarth Objects (NEOs). As we try to safeguard our planet from capacity effect activities, proactive steps are taken to minimize the risks and ensure a wellprepared reaction. Here are the key additives of precautionary and preparedness measures:

Early Detection and Monitoring: Continuous tracking of the skies using superior telescopes and radar structures is critical for early detection of PHAs. Identifying those celestial our bodies as early as feasible lets in for sufficient time to evaluate their trajectories and potential effect dangers.

Impact Risk Assessment and Characterization: Sophisticated fashions and simulations are employed to evaluate the effect risks of PHAs correctly. Detailed characterizations of asteroid size, composition, and capability impact effects are critical for making informed selections.

Planetary Defense Exercises and Drills: Regular physical games and drills are conducted to simulate capability effect scenarios and check response plans. These physical games contain relevant stakeholders and assist refine emergency preparedness protocols.

Emergency Response and Communication: Governments and government increase comprehensive emergency reaction plans that outline communique strategies, evacuation tactics, and coordination amongst businesses. Effective conversation in the course of a capacity impact occasion is vital for public protection.

International Collaboration: Planetary protection is a global effort, and international collaboration is essential for sharing information, understanding, and assets. Cooperation amongst nations allows a unified reaction to potential impact threats.

Spacecraft and Satellite Safety: As we increase human presence in space, it's far critical to ensure the safety of spacecraft and satellites. Implementing measures to defend those belongings from capacity asteroid impacts is essential for area exploration endeavors.

Asteroid Characterization and Exploration: Closeup exploration of asteroids through area missions affords valuable information about their composition and traits. This statistics enhances our know-how of capacity effect dangers and informs mitigation strategies.

Public Awareness and Education: Raising public attention about asteroid threats and planetary protection efforts is essential. Educational applications, media campaigns, and outreach tasks foster a sense of duty and preparedness among the overall public.

International Protocols and Policies: Governments and space groups paintings together to establish international protocols and regulations for planetary protection. These frameworks address problems consisting of statistics sharing, coordination, and response mechanisms.

Research and Technology Development: Investing in research and technology development is vital for advancing our capabilities in planetary protection. Innovations in area commentary, deflection techniques, and impact modeling make contributions to higher preparedness.

Precautionary and preparedness measures are crucial in mitigating the ability risks posed through Potentially Hazardous Asteroids and other NearEarth Objects. Early detection, impact danger assessment, emergency preparedness, and international collaboration are key additives of our efforts to shield our planet and make certain a safer destiny for humanity. Through continuous studies, training, and technological advancements, we are able to enhance our resilience and readiness to cope with ability impact activities efficaciously.

10.2. Recommendations for the Future

International collaboration and records sharing are important factors in the area of planetary protection, in particular in terms of addressing capability threats posed by Potentially Hazardous Asteroids (PHAs) and other NearEarth Objects (NEOs). The vastness of space and the global nature of asteroid effect dangers necessitate cooperation and information change amongst countries and area groups. Here are the key factors of worldwide collaboration and data sharing in planetary protection:

Global Information Network: Astronomical observations and data amassed from diverse telescopes and radar structures round the arena shape a international facts community. This network enables the sharing of important observational statistics, allowing for complete monitoring of potential PHAs and NEOs.

Asteroid FollowUp Observations: Once a new PHA or NEO is detected, followup observations are vital to refine its orbit and examine capacity effect risks correctly. International collaboration guarantees that multiple observatories worldwide take part within the followup efforts, enhancing the accuracy of the information.

Coordination of Space Missions: Space missions aimed toward reading asteroids or engaging in deflection assessments regularly involve a couple of international locations and area agencies. International collaboration guarantees clean coordination amongst special missions and optimizes useful resource usage.

Joint Research and Analysis: Researchers and scientists from numerous nations collaborate on joint studies and evaluation of asteroid records. Pooling information and sources result in extra complete research, leading to a better know-how of asteroid characteristics and behaviors.

Sharing Impact Risk Assessments: Impact hazard checks for PHAs are essential in figuring out the ability threat they pose to Earth. By sharing those tests globally, international locations can collectively put together and plan for capacity effect eventualities.

Space Situational Awareness: International collaboration in area situational attention allows track asteroids and space particles, minimizing the risks of collisions with spacecraft and satellites. This shared data supports safe space exploration and satellite operations.

Information Exchange Platforms: Dedicated statistics trade platforms and databases facilitate the sharing of asteroidrelated information amongst international companions. These structures streamline verbal exchange and make sure timely access to important information.

Disaster Preparedness and Response: In the event of a capacity asteroid impact chance, global collaboration performs a pivotal position in disaster preparedness and reaction. Sharing impact predictions and reaction techniques allows international locations coordinate efforts to guard human lives and infrastructure.

Standardization of Data Formats: To permit seamless facts sharing and integration, global agreements regularly promote standardization of statistics codecs and terminology. Consistent facts codecs decorate information compatibility and facilitate collaborative research.

Policy and Legal Frameworks: International collaboration in planetary protection requires coverage and prison frameworks that promote statistics sharing and cooperation. Bilateral and multilateral agreements lay the inspiration for powerful collaboration in this critical discipline.

International collaboration and statistics sharing are vital pillars of planetary protection efforts. Working collectively as a worldwide community allows us to monitor and cope with capability threats from Potentially Hazardous Asteroids and NearEarth Objects greater efficiently. Through shared records, joint research, and coordinated efforts, we are able to beautify our preparedness and reaction abilties, making sure a safer destiny for our planet and humanity as we navigate the cosmos.

Future research and missions play a pivotal function in advancing our know-how of Potentially Hazardous Asteroids (PHAs) and enhancing our preparedness for capacity impact occasions. As technology and area exploration hold to evolve, the following factors are vital in shaping future studies and missions within the field of planetary protection:

Advanced Observation Techniques: Developing and deploying extra advanced telescopes, spacebased observatories, and radar systems will permit extra unique and complete observations of PHAs. These advanced statement strategies will enhance our capacity to stumble on, song, and symbolize capacity threats.

Asteroid Characterization: Future missions will consciousness on closeup exploration of asteroids to have a look at their bodily homes, composition, and internal structures. Understanding these traits is crucial for devising effective deflection strategies and impact threat checks.

Mitigation Technology Testing: Research missions will take a look at and validate diverse deflection and mitigation technologies in managed environments. This will improve our self assurance in enforcing those techniques while facing actual capacity impact scenarios.

Collaboration and Data Sharing: Continued global collaboration and facts sharing among area groups, studies establishments, and astronomers are vital for growing a global network of records to screen and manipulate PHAs correctly.

NEO Survey Missions: Future space missions will focus on carrying out comprehensive surveys of NearEarth Objects to identify and categorize capability effect threats. These surveys will assist prioritize and allocate resources for planetary protection efforts.

Orbit Determination and Prediction: Advancements in computational modeling and synthetic intelligence will permit more accurate and green orbit determination and prediction of PHAs, decreasing uncertainties in effect danger tests.

Planetary Defense Testing: Future studies missions will conduct planetary defense testing by way of simulating asteroid impact situations and studying the effectiveness of deflection techniques. These exams will tell and refine our mitigation techniques.

Space Situational Awareness: Continued funding in area situational attention and space site visitors management will beautify our potential to screen and expect capacity asteroid collisions with spacecraft and satellites.

Public Awareness and Education: Future research and missions will vicinity extended emphasis on public recognition and training approximately asteroid threats and planetary protection efforts. Educating the general public will foster aid and expertise of the importance of these endeavors.

Planetary Defense Policy and International Cooperation: Developing strong coverage frameworks and promoting international cooperation will make sure a coordinated and unified response to potential effect threats.

Future studies and missions in planetary defense are driven by way of improvements in remark generation, characterization of asteroids, and checking out of mitigation techniques. Collaboration, data sharing, and space situational consciousness will strengthen our collective efforts in safeguarding our planet from capacity asteroid influences. Public cognizance and global cooperation are crucial to ensure a more secure future for humanity as we hold to explore and defend our celestial community.

10.3. Promising Developments and Resources

In latest years, significant development has been made inside the field of planetary defense, with promising trends and assets rising to address ability asteroid threats. These improvements and assets are instrumental in enhancing our preparedness and response talents. Here are some of the important thing promising tendencies and resources:

Advanced Space Telescopes: Stateoftheart area telescopes equipped with superior sensors and imaging abilties have revolutionized our potential to locate and track NearEarth Objects (NEOs) and Potentially Hazardous Asteroids (PHAs). These telescopes offer highresolution pix and unique facts, improving our know-how of the traits and trajectories of those celestial items.

SpaceBased Radar Systems: Spacebased radar structures complement groundbased observations by way of presenting treasured records on asteroids' size, form, rotation, and floor features. This technology enhances our capacity to assess capacity impact risks and broaden powerful deflection techniques.

NearEarth Asteroid Missions: Space missions centered at reading nearEarth asteroids up close have provided treasured insights into their composition, structure, and ability threat level. These missions, which includes NASA's OSIRISREx and

Japan's Hayabusa2, have accumulated samples from asteroids, offering important data for scientific analysis.

Planetary Defense Coordination Office: International area businesses, together with NASA, have hooked up Planetary Defense Coordination Offices (PDCOs) to centralize efforts in tracking, assessing, and responding to potential effect threats. PDCOs sell international collaboration and facilitate facts alternate among countries.

International Asteroid Warning Network (IAWN): The IAWN is a worldwide network of observatories and establishments centered on detecting and tracking NEOs and PHAs. This collaborative effort ensures well timed facts sharing and worldwide coordination inside the face of capacity impact activities.

Asteroid Impact Deflection Assessment (AIDA) Mission: A joint NASAESA undertaking, AIDA pursuits to look at the effectiveness of asteroid deflection strategies. The undertaking entails the Double Asteroid Redirection Test (DART) spacecraft impacting the asteroid Dimorphos, even as the Hera spacecraft observes the effects.

Space Resources for Asteroid Deflection: Innovative proposals explore the usage of area sources, along with mining asteroids for treasured materials, which will be utilized in asteroid deflection techniques. These efforts may want to result in more sustainable and costeffective deflection missions.

Planetary Defense Technology Development: Advances in propulsion structures, nanosatellites, and highpower lasers provide potential technology for asteroid deflection and impact threat mitigation. Continued research and improvement in these areas preserve promise for future planetary defense missions.

PublicPrivate Partnerships: Collaboration among governmental space agencies and private area groups has led to multiplied investment and innovation in planetary defense. Publicprivate partnerships boost up technological advancements and increase assets to be had for studies and missions.

Planetary Defense Awareness Campaigns: Governments and area companies are actively attractive the public via awareness campaigns, educational applications, and citizen science initiatives. These efforts foster public help, expertise, and engagement in planetary protection endeavors.

Promising trends and resources in planetary defense have substantially advanced our capabilities to detect, tune, and mitigate capacity asteroid threats. From advanced area telescopes and radar systems to asteroid missions and worldwide collaboration networks, these tendencies offer the muse for a more secure future. As era maintains to evolve and worldwide cooperation strengthens, our capability to shield our planet from potential asteroid affects will retain to enhance, making sure the renovation of lifestyles on Earth.

Printed in Dunstable, United Kingdom